职业教育国家在线精品课程配套教材

Python 大数据技术系列

数据采集技术
——Python网络爬虫项目化教程（第2版）

黄锐军　主编

中国教育出版传媒集团

高等教育出版社·北京

内容简介

网络爬虫是一种能自动从网站的相关网页中搜索与提取数据的程序或脚本，采集这些数据是进一步实现数据分析的前提与关键。Python 语言语法简洁、开发效率高，用于编写网络爬虫有特别的优势，尤其业界有专门为 Python 语言编写的各种各样的爬虫程序框架，使得使用 Python 语言编写爬虫程序更加便捷、高效。

本书包括 5 个项目：项目 1 为爬取学生信息，讲解了如何使用 Python 访问 Web，该技术是编写爬虫程序的基础；项目 2 为爬取天气预报数据，讲解了网页数据的爬取方法，其中重点讲解了如何使用 BeautifulSoup 进行数据采集与数据分析；项目 3 为爬取旅游网站数据，讲解了爬取多个网页数据的方法，重点讲解了如何使用深度优先与广度优先策略爬取数据，同时还讲解了如何使用多线程爬取网页数据；项目 4 为爬取航空网站数据，讲解了目前功能强大的分布式爬取框架 Scrapy；项目 5 为爬取商城网站数据，讲解了 Selenium 动态网页数据的爬取技术。每个项目都遵循由浅入深的学习规律，理论与实践相结合，实现了实用的爬虫程序，进而提高读者的实践能力。

本书为新形态一体化教材，配有丰富的教学资源，包括微课、教学大纲、课程标准、教学课件、案例源码、课后习题及习题答案等。与本书配套的数字课程在"智慧职教"平台（www.icve.com.cn）上线，学习者可以登录平台进行在线学习，授课教师可以调用本课程构建符合自身教学特色的 SPOC 课程，详见"智慧职教"服务指南。教师也可发邮件至编辑邮箱 1548103297@qq.com 获取相关教学资源。

本书可作为高等职业教育大数据技术、人工智能技术应用、软件技术及其他计算机类专业的数据采集类课程教材，也可作为数据采集技术学习者的自学参考书。

图书在版编目（CIP）数据

数据采集技术：Python 网络爬虫项目化教程 / 黄锐军主编. --2 版. --北京：高等教育出版社，2023.4
ISBN 978-7-04-059690-8

Ⅰ.①数… Ⅱ.①黄… Ⅲ.①软件工具-程序设计-高等职业教育-教材 Ⅳ.①TP311.561

中国国家版本馆 CIP 数据核字（2023）第 006723 号

SHUJU CAIJI JISHU——PYTHON WANGLUO PACHONG XIANGMUHUA JIAOCHENG

策划编辑	白 颢	责任编辑	侯昀佳	封面设计	姜 磊	版式设计	于 婕
责任绘图	李沛蓉	责任校对	刘丽娴	责任印制	存 怡		

出版发行	高等教育出版社		网　　址	http://www.hep.edu.cn
社　　址	北京市西城区德外大街 4 号			http://www.hep.com.cn
邮政编码	100120		网上订购	http://www.hepmall.com.cn
印　　刷	北京利丰雅高长城印刷有限公司			http://www.hepmall.com
开　　本	787 mm×1092 mm　1/16			http://www.hepmall.cn
印　　张	14.5		版　　次	2018 年 8 月第 1 版
字　　数	360 千字			2023 年 4 月第 2 版
购书热线	010-58581118		印　　次	2023 年 4 月第 1 次印刷
咨询电话	400-810-0598		定　　价	43.00 元

本书如有缺页、倒页、脱页等质量问题，请到所购图书销售部门联系调换
版权所有　侵权必究
物 料 号　59690-00

"智慧职教"服务指南

"智慧职教"（www.icve.com.cn）是由高等教育出版社建设和运营的职业教育数字教学资源共建共享平台和在线课程教学服务平台，与教材配套课程相关的部分包括资源库平台、职教云平台和 App 等。用户通过平台注册，登录即可使用该平台。

● 资源库平台：为学习者提供本教材配套课程及资源的浏览服务。

登录"智慧职教"平台，在首页搜索框中搜索"数据采集技术——Python 网络爬虫项目化教程"，找到对应作者主持的课程，加入课程参加学习，即可浏览课程资源。

● 职教云平台：帮助任课教师对本教材配套课程进行引用、修改，再发布为个性化课程（SPOC）。

1. 登录职教云平台，在首页单击"新增课程"按钮，根据提示设置要构建的个性化课程的基本信息。

2. 进入课程编辑页面设置教学班级后，在"教学管理"的"教学设计"中"导入"教材配套课程，可根据教学需要进行修改，再发布为个性化课程。

● App：帮助任课教师和学生基于新构建的个性化课程开展线上线下混合式、智能化教与学。

1. 在应用市场搜索"智慧职教 icve" App，下载安装。

2. 登录 App，任课教师指导学生加入个性化课程，并利用 App 提供的各类功能，开展课前、课中、课后的教学互动，构建智慧课堂。

"智慧职教"使用帮助及常见问题解答请访问 help.icve.com.cn。

第 2 版前言

网络爬虫是一种能自动从网站的相关网页中自动搜索与提取数据的程序，提取与存储这些数据是进一步实现数据分析的前提与关键。Python 语言是一种面向对象的解释型计算机程序设计语言。该语言开源、免费、功能强大，而且语法简洁清晰，具有丰富和强大的库，是目前应用广泛的程序设计语言。使用 Python 语言编写网络爬虫有特别的优势，尤其业界有专门为 Python 语言编写的各种各样的爬虫程序框架，使得爬虫程序的编写更加便捷高效。

本书第 1 版出版后，基于广大院校师生的教学应用反馈并结合最新的课程教学改革成果，不断优化、更新内容，以将最新的爬虫技术、理念以及行业发展动态纳入教学。此外，为加快推进党的二十大精神进教材、进课堂、进头脑，在本版修订过程中，优化修改了部分项目的教学案例，并增加了动态数据爬取等教学内容。体现了现代信息技术与教育教学的深度融合，进一步推动教育数字化发展。

本书包括 5 个项目：项目 1 为爬取学生信息，讲解了如何使用 Python 访问 Web，该技术是编写爬虫程序的基础；项目 2 为爬取天气预报数据，讲解了网页数据的爬取方法，其中重点讲解了如何使用 BeautifulSoup 进行数据采集与数据分析；项目 3 为爬取旅游网站数据，讲解爬取多个网页数据的方法，重点讲解了如何使用深度优先与广度优先策略爬取数据，同时还讲解了如何使用多线程爬取网页数据；项目 4 为爬取航空网站数据，讲解了目前功能强大的分布式爬取框架 Scrapy；项目 5 为爬取商城网站数据，讲解了 Selenium 动态网页数据的爬取技术。每个项目都遵循由浅入深的学习规律，理论与实践相结合，实现了实用的爬虫程序。

本书可作为高等职业教育软件技术、大数据技术、人工智能技术应用及其他计算机类专业的教材，也可作为数据采集技术学习者的自学参考书。

由于编者知识水平有限，书中难免有疏漏和错误之处，欢迎广大读者批评指正。

黄锐军
2023 年 1 月

目录

项目 1　爬取学生信息　1

1.1　爬虫程序开发环境　2
　1.1.1　爬虫程序简介　2
　1.1.2　搭建 Python 开发环境　2
1.2　Flask Web 网站　3
　1.2.1　Flask 简介　3
　1.2.2　调用 Urllib 库访问 Web 网站　6
1.3　使用 GET 方法访问网站　7
　1.3.1　客户端使用 GET 方法发送数据　7
　1.3.2　服务器获取 GET 方法发送的数据　7
1.4　使用 POST 方法向网站发送数据　9
　1.4.1　客户端使用 POST 方法发送数据　9
　1.4.2　服务器获取 POST 方法的数据　10
　1.4.3　GET 与 POST 方法的混合使用　10
1.5　搭建图书网站　12
　1.5.1　准备网站素材　13
　1.5.2　创建网页模板　13
　1.5.3　创建网站服务器程序　15
　1.5.4　运行网站服务器程序　16
1.6　正则表达式与查找匹配字符串　16
　1.6.1　正则表达式　16
　1.6.2　查找匹配字符串　21
1.7　实践项目——爬取学生信息　22
　1.7.1　项目简介　22
　1.7.2　服务器程序　22
　1.7.3　客户端程序　23
练习一　27

项目 2　爬取天气预报数据　29

2.1　HTML 文档结构与文档树　30
　2.1.1　HTML 文档结构　30
　2.1.2　HTML 文档树　31
2.2　BeautifulSoup 装载 HTML 文档　31
　2.2.1　BeautifulSoup 的安装　31
　2.2.2　BeautifulSoup 装载 HTML 文档　31
2.3　查找文档元素　35
　2.3.1　查找 HTML 元素　35
　2.3.2　获取 HTML 元素属性值　38
　2.3.3　获取元素包含的文本值　39
　2.3.4　高级查找　41
2.4　遍历文档元素　43
　2.4.1　获取元素节点的父节点元素　43
　2.4.2　获取元素节点的所有子节点元素　44
　2.4.3　获取元素节点的所有子孙节点元素　44
　2.4.4　获取元素节点的兄弟节点　45
2.5　使用 CSS 语法查找元素　46
　2.5.1　使用 CSS 语法　46
　2.5.2　属性的语法规则　48
　2.5.3　使用 soup.select()查找子孙节点　48
　2.5.4　使用 soup.select()查找直接子节点　49
　2.5.5　使用 soup.select()查找兄弟节点　49
2.6　爬取图书网站数据　50
　2.6.1　分析网站结构　50
　2.6.2　获取图书数据　50
　2.6.3　编写爬虫程序　51
2.7　实践项目——爬取天气预报数据　54
　2.7.1　项目简介　54
　2.7.2　HTML 代码分析　54
　2.7.3　爬取天气预报数据　58
　2.7.4　爬取与存储天气预报数据　59
练习二　62

项目3 爬取旅游网站数据 ... 65

- 3.1 网站树的爬取路径 ... 66
  - 3.1.1 Web 服务器网站 ... 66
  - 3.1.2 使用递归程序爬取数据 ... 68
  - 3.1.3 使用深度优先策略爬取数据 ... 69
  - 3.1.4 广度优先策略爬取数据 ... 70
- 3.2 网站图的爬取路径 ... 72
  - 3.2.1 复杂的 Web 网站 ... 72
  - 3.2.2 改进客户端深度优先策略程序 ... 73
  - 3.2.3 改进客户端广度优先策略程序 ... 75
- 3.3 Python 实现多线程 ... 77
  - 3.3.1 Python 的前后台线程 ... 77
  - 3.3.2 线程的等待 ... 79
  - 3.3.3 多线程与资源 ... 81
- 3.4 爬取网站复杂数据 ... 83
  - 3.4.1 Web 服务器网站 ... 83
  - 3.4.2 爬取网站的复杂数据 ... 84
  - 3.4.3 爬取程序的改进 ... 86
- 3.5 爬取网站的图像文件 ... 89
  - 3.5.1 项目简介 ... 89
  - 3.5.2 单线程爬取图像的程序 ... 90
  - 3.5.3 多线程爬取图像的程序 ... 91
- 3.6 爬取图书网站数据 ... 93
  - 3.6.1 分析网站结构 ... 93
  - 3.6.2 换页递归爬取 ... 94
  - 3.6.3 图书数据存储 ... 94
  - 3.6.4 编写爬虫程序 ... 96
  - 3.6.5 执行爬虫程序 ... 98
- 3.7 实践项目——爬取旅游网站数据 ... 99
  - 3.7.1 实践项目简介 ... 99
  - 3.7.2 网站网页分析 ... 99
  - 3.7.3 网站数据爬取 ... 101
  - 3.7.4 网站网页翻页 ... 101
  - 3.7.5 网站数据存储 ... 102
  - 3.7.6 编写爬虫程序 ... 102
  - 3.7.7 执行爬虫程序 ... 107
- 练习三 ... 108

项目4 爬取航空网站数据 ... 109

- 4.1 Scrapy 框架爬虫简介 ... 110
  - 4.1.1 安装 Scrapy 框架 ... 110
  - 4.1.2 建立 Scrapy 项目 ... 110
  - 4.1.3 入口函数与入口地址 ... 113
  - 4.1.4 Python 的 yield 语句 ... 113
- 4.2 Scrapy 中查找 HTML 元素 ... 114
  - 4.2.1 Scrapy 的 XPath 简介 ... 114
  - 4.2.2 XPath 查找 HTML 元素 ... 116
- 4.3 Scrapy 爬取与存储数据 ... 126
  - 4.3.1 建立 Web 网站 ... 126
  - 4.3.2 编写数据项目类 ... 127
  - 4.3.3 编写爬虫程序 mySpider.py ... 128
  - 4.3.4 编写数据管道处理类 ... 129
  - 4.3.5 设置 Scrapy 的配置文件 ... 130
- 4.4 Scrapy 爬取网站数据 ... 130
  - 4.4.1 建立 Web 网站 ... 131
  - 4.4.2 编写 Scrapy 爬虫程序 ... 132
  - 4.4.3 存储 Scrapy 爬取的数据 ... 134
- 4.5 实践项目——爬取图书网站数据 ... 135
  - 4.5.1 网站结构分析 ... 136
  - 4.5.2 图书数据爬取 ... 136
  - 4.5.3 图书数据存储 ... 137
  - 4.5.4 设计爬虫程序 ... 138
  - 4.5.5 执行爬虫程序 ... 142
- 4.6 实践项目——爬取航空网站数据 ... 142
  - 4.6.1 项目简介 ... 142
  - 4.6.2 网页结构分析 ... 142
  - 4.6.3 爬取航班数据 ... 146
  - 4.6.4 获取换页地址 ... 147
  - 4.6.5 存储航班数据 ... 148
  - 4.6.6 编写爬虫程序 ... 148
  - 4.6.7 执行爬虫程序 ... 152
- 练习四 ... 153

项目5 爬取商城网站数据 ... 157

- 5.1 使用 Selenium 编写爬虫程序 ... 158
  - 5.1.1 JavaScript 控制网页 ... 158

5.1.2	普通爬虫程序问题	159
5.1.3	安装 Selenium 框架	160
5.1.4	编写 Selenium 爬虫程序	161
5.2	使用 Selenium 查找 HTML 元素	163
5.2.1	创建产品网站	163
5.2.2	使用 Xpath 查找元素	164
5.2.3	查找元素的文本与属性	165
5.2.4	使用 id 查找元素	167
5.2.5	使用 name 查找元素	167
5.2.6	使用 CSS 查找元素	168
5.2.7	使用标签查找元素	169
5.2.8	查找超链接	169
5.2.9	使用 class 查找元素	170
5.3	使用 Selenium 实现用户登录	171
5.3.1	创建用户登录网站	171
5.3.2	使用元素动作	172
5.3.3	编写爬虫程序	174
5.3.4	执行 JavaScript 程序	175
5.4	使用 Selenium 爬取 Ajax 网页数据	176
5.4.1	创建 Ajax 网站	177
5.4.2	理解 Selenium 爬虫程序	179
5.4.3	编写爬虫程序	181
5.4.4	执行爬虫程序	183
5.5	爬取网站换页数据	183
5.5.1	创建实验网站	183
5.5.2	爬虫程序问题	185
5.5.3	编写爬虫程序	185
5.5.4	执行爬虫程序	186
5.6	使用 Selenium 等待 HTML 元素	186
5.6.1	创建延迟网站	186
5.6.2	编写爬虫程序	188
5.6.3	Selenium 强制等待	189
5.6.4	Selenium 隐式等待	189
5.6.5	Selenium 显式等待	190
5.6.6	Selenium 等待形式	192
5.7	爬取图书网站数据	193
5.7.1	网站结构分析	193
5.7.2	获取网站数据	193
5.7.3	图书数据存储	194
5.7.4	编写爬虫程序	195
5.8	实践项目——爬取商城网站数据	199
5.8.1	解析网页代码	199
5.8.2	爬取网页数据	203
5.8.3	实现网页翻页	204
5.8.4	商品数据存储	206
5.8.5	编写爬虫程序	207
5.8.6	执行爬虫程序	213
练习五		214
结语		217
参考文献		219

项目 1　爬取学生信息

本项目介绍使用 Flask 微型框架快速部署 Web 服务的方法，以及使用 Urllib 库访问网站的过程，最后通过完成一个学生记录管理程序的综合案例来巩固知识与技能。

PPT 爬虫程序开发环境

微课 1
爬虫程序开发环境

1.1 爬虫程序开发环境

1.1.1 爬虫程序简介

爬虫程序是一种部署在客户端上的程序,功能是通过访问 Web 服务器,从服务器中获取网页代码,网页代码中包含了很多数据信息,爬虫程序从中提取需要的数据,将数据整理后存储在本地的数据库中,这些数据可以被应用在数据分析等模块中。

例如,想知道城市的天气预报,可以设计程序访问有天气预报数据的网站,天气预报的网页代码,如图 1-1-1 所示。爬虫程序的任务就是要从这些复杂的代码中提取需要的天气状况、温度、风力等数据,并把这些数据存储在数据库中。

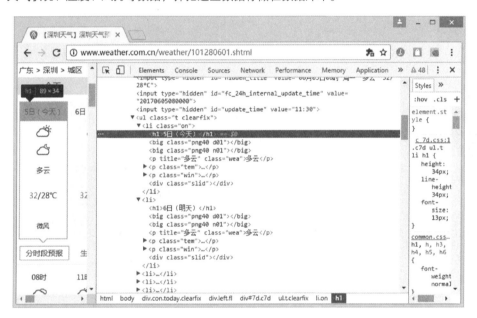

图 1-1-1
网页代码

编写爬虫程序可以使用 Python、Java、C++、C#等开发语言,但是使用 Python 语言编写比较简单也比较流行。

爬虫程序爬取的数据量很大,但相关的数据一般分布在很多不同的网页中,甚至分布在相关联的不同的网站中爬虫程序必须能按链接自动在这些不同的网站中爬取数据,一个爬虫程序爬取百万、千万条数据是常有的事,怎样设计一个高效率的爬虫程序是本书的重点。

1.1.2 搭建 Python 开发环境

Python 是一种面向对象的解释型计算机程序设计语言,于 1989 年发明,第一个公开发行版于 1991 年发行。

Python 语言具有以下特点:

- 免费、开源、功能强大。
- 语法简洁清晰,强制用空白符(White Space)作为语句缩进。
- 具有丰富和强大的库。
- 易读、易维护、受欢迎且用途广泛。

- 解释型语言、变量类型可变，类似 JavaScript。

Python 安装后带有命令行工具与 IDE 程序，但是该 IDE 程序功能很弱，推荐搭配第三方的 IDE 开发工具，主流的开发工具与环境如下所示。

（1）Python 官方开发环境

Python 的官方开发环境十分轻量，用户可以到 Python 官方网站上直接下载 Python 的程序包。目前 Python 有两个主流的版本：一个是 Python 2.7 版本；另一个是 Python 3.6 版本。这两个版本在语法上有差异，本书使用 Python 3.6 版本。

下载 Python 3.6 程序包后，选择安装目录直接安装。Python 安装完毕后在 Windows 启动菜单中可以看到 Python 3.6 的启动项，启动 Python 3.6 可以看到 Python 的命令行界面。这个环境是命令行环境，适合编写并运行简单的测试语句，不适合编写大型程序。Python 官方自带的 IDE 的功能有限，不适合开发 Python 工程项目。

（2）PyCharm

PyCharm 是 Python 常用的 IDE，风格和 Eclipse 类似，带有可以帮助用户在使用 Python 语言开发时提高其效率的工具库，比如调试、语法高亮、项目管理、代码跳转、智能提示、自动补全、单元测试、版本控制等。

从 PyCharm 官网可以下载免费的 PyCharm Community 版本。

（3）Anaconda

Anaconda 是一个十分强大的 Python 开发环境。安装 Anaconda 时会自动安装指定版本的 Python，同时还带有功能强大的 IDE 开发工具——Spider。Anaconda 可以进行版本控制，在不同 Python 环境中切换，还可以帮助用户查询与安装 Python 的库，使得 Python 开发十分方便与高效。

1.2　Flask Web 网站

1.2.1　Flask 简介

Python 的 Web 程序开发框架很多，Flask 是一个非常容易上手的 Python Web 开发框架。使用该框架不需要知道太多的 MVC 的概念，只需要具备基本的 Python 开发技能，就可以开发出一个 Web 应用。

Flask 的官方网站：https://flask.net.cn。

推荐先阅读官方网站中的"安装"文档，然后阅读"快速上手"文档。Flask 具有强大的扩展功能，支持以各种方式扩展 Flask 的功能，比如增强对数据库的支持等。

（1）安装 Flask

在 Windows 系统中使用 Flask，安装方法非常简单，根据文档的介绍直接在命令行窗口执行：

```
pip install flask
```

如果最后显示：

```
Successfully installed flask Werkzeug Jinja2 itsdangerous markupsafe
Cleaning up...
```

表示 Flask 已安装成功。

（2）Flask 实例

编写程序：

```
import flask
app=flask.Flask(__name__)


@app.route("/")
def hello():
    return "你好"


@app.route("/hi")
def hi():
    return "Hi,你好"


if __name__=="__main__":
    app.run()
```

执行该程序可以看到显示 http://127.0.0.1:5000 的 Web 地址，在浏览器中输入该 Web 地址看到显示"你好"，如果输入 http://127.0.0.1:5000/hi 则显示"Hi,你好"。

下面分析程序的功能：

```
import flask
```

这条语句是导入 Flask 库，在 Flask 正确安装后支持正常导入。

```
app=flask.Flask(__name__)
```

这条语句是初始化一个 Flask 对象，参数__name__是程序的名称。

```
@app.route("/")
def hello():
    return "你好"
```

这是一段路由控制语句，每个路由地址用@app.route(...)方法中的参数来指明，在访问相对地址是"/"时执行函数 hello()，因此在访问地址 http://127.0.0.1:5000 时显示"你好"。

```
@app.route("/hi")
def hi():
    return "Hi,你好"
```

这也是一段路由控制语句，在访问相对地址/hi 时执行函数 hi()，因此访问地址 http://127.0.0.1:5000/hi 时显示"Hi,你好"。

```
if __name__=="__main__":
    app.run()
```

这一段语句表示在主程序中执行 app.run()，一旦执行 app.run()后就启动了一个 Web 服务器，默认地址是http://127.0.0.1:5000。

（3）Flask 显示静态网页

如果在程序的同一文件夹中有一个静态网页，例如 index.htm，那么很容易使用 Flask 编写一个 Web 程序 server.py，它的主页就是 index.htm，具体程序如下：

```
import flask
app=flask.Flask(__name__)

@app.route("/")
def index():
    try:
        fobj=open("index.htm","rb")
        data=fobj.read()
        fobj.close()
        return data
    except Exception as err:
        return str(err)

if __name__=="__main__":
    app.run()
```

程序 server.py 的功能是启动一个 Web 服务，在访问网站时读取同一个文件夹下的 index.htm 文件，然后向客户端（浏览器）返回 index.htm 文件的内容。

例如 index.htm 的内容是：

```
<h1>Welcome Python Flask Web</h1>
It is very easy to make a website by Python Flask.
```

设置该文件的编码格式为 UTF-8 并保存到 Python 程序 service.py 所在的同级文件夹中，运行程序后访问网址 http://127.0.0.1:5000，如图 1-2-1 所示。

图 1-2-1
Flask Web 网站

1.2.2 调用 Urllib 库访问 Web 网站

微课 3
Urllib 程序包访问 Web 网站

server.py 程序生成的网站除了可以使用浏览器访问，也可以调用 Urllib 库中的相关函数来访问。设计 client.py 程序如下：

```
import urllib.request
url="http://127.0.0.1:5000"
html = urllib.request.urlopen(url)
html = html.read()
html = html.decode()
print(html)
```

运行 server.py 程序后运行 client.py，显示结果如下：

```
<h1>欢迎使用 Python Flask Web</h1>
It is very easy to make a website by Python Flask.<p>
我们很容易用 Python Flask 制作一个 Web 网站<p>
```

显示的是 index.htm 网页的内容，下面来分析这个程序的功能。

```
import urllib.request
```

这条语句的作用是导入 urllib.request 库，该库是 Python 自带的库，不需要单独安装，其作用是访问网站。

```
html = urllib.request.urlopen(url)
```

这条语句的作用是打开参数 url 的网址，其中 urllib.request 是 Urllib 库中的一个子库，urlopen()是打开网站的函数。

```
html = html.read()
```

网站打开后就如同打开文件一样，要使用 read()函数读取网站的内容，返回的数据是二进制数据。

```
html = html.decode()
```

这条语句的作用是把二进制数据 html 解码为字符串，默认文件的编码是 UTF-8，decode()默认也是使用 UTF-8 编码，也可以指定编码，例如 html=html.decode("utf-8")或 html=html.decode("gbk")，具体使用什么编码需要依据网站的网页说明，如果编码不正确会出现汉字乱码的现象。

```
print(html)
```

显示网站的网页内容，可见显示的就是 index.htm 的网页数据。
urllib.request.urlopen(url)是一个很重要的函数，可以打开一个 URL 网址的网站。

1.3 使用 GET 方法访问网站

PPT　GET 方法访问网站

微课 4
GET 方法访问网站

访问网站最常用的一种方法是 GET 方法，这种方法主要是客户端从服务器获取网站数据，如有交互，客户端通过把参数附加在网址的后面向服务器提供参数，服务器接收参数并响应。

微课 5
客户端 GET 方式发送数据

1.3.1 客户端使用 GET 方法发送数据

GET 方法发送需要将数据附加在 URL 后面，在 URL 后面先接一个?号，数据采用"名称 1=值 1&名称 2=值 2&名称 3=值 3…"的方式，多个数据之间用&符号隔开。例如，向服务器传递省份与城市的数据就可以这样写：

```
urllib.request.urlopen("http://127.0.0.1:5000?province=GD&city=SZ")
```

如果参数值包含汉字，那么需要使用 urllib.parse.quote()方法对参数值进行编码，例如：

```
province= urllib.parse.qoute("广东")
city= urllib.parse.qoute("深圳")
urllib.request.urlopen("http://127.0.0.1:5000?province="+province+"&city="+city)
```

编写客户端程序 client.py 如下：

```
import urllib.parse
import urllib.request
url="http://127.0.0.1:5000"
try:
    province= urllib.parse.quote("广东")
    city= urllib.parse.quote("深圳")
    data="province="+province+"&city="+city
    html=urllib.request.urlopen("http://127.0.0.1:5000?"+data)
    html = html.read()
    html = html.decode()
    print(html)
except Exception as err:
    print(err)
```

1.3.2 服务器获取 GET 方法发送的数据

服务器用 Flask 库的 request 对象中的 args 来存储 GET 的参数，使用 GET 方法来获取参数，即用 flask.request.args.get（参数）来获取参数的值，例如：

```
province=flask.request.args.get("province")
city=flask.request.args.get("city")
```

就可以获取 GET 传递的参数 province 与 city 的值。

编写服务器程序 server.py 如下：

```python
import flask
app=flask.Flask(__name__)

@app.route("/")
def index():
    try:
        province=flask.request.args.get("province")
        city = flask.request.args.get("city")
        return province+","+city
    except Exception as err:
        return str(err)

if __name__=="__main__":
    app.run()
```

先运行 server.py 建立 Web 网站，再运行 client.py，可以看到 client.py 程序运行结果：

```
广东,深圳
```

注意 Web 网址是 http://127.0.0.1:5000，如果直接访问这个网站会导致服务器程序报错，因为访问时没有提供 province 与 city 参数，服务器在执行如下语句时会到 args 中查找 province 和 city 参数的值，结果没有找到而出现错误。

```python
province=flask.request.args.get("province")
city = flask.request.args.get("city")
```

为了避免这种错误可以把这两条语句优化成如下语句：

```python
province=flask.request.args.get("province") if "province" in flask.request.args else ""
city = flask.request.args.get("city") if "city" in flask.request.args else ""
```

这样在 province、city 没有出现在 flask.request.args 中时就设置 province、city 值为空字符串。服务器程序整体优化如下：

```python
import flask
app=flask.Flask(__name__)

@app.route("/")
def index():
    try:
        province=flask.request.args.get("province") if "province" in flask.request.args else ""
        city = flask.request.args.get("city") if "city" in flask.request.args else ""
        return province+","+city
```

```
        except Exception as err:
            return str(err)


if __name__=="__main__":
    app.run()
```

1.4 使用 POST 方法向网站发送数据

1.4.1 客户端使用 POST 方法发送数据

使用 POST 方法访问网站时，客户端向服务器发送表单数据，表单数据的组织方式与 GET 方法的参数列表十分相似，结构如下：

名称 1=值 1&名称 2=值 2&名称 3=值 3…

多个数据之间用"&"符号隔开，如果参数值包含汉字，那么需要 urllib.parse.quote() 方法对参数值进行编码，例如：

```
province= urllib.parse.qoute("广东")
city= urllib.parse.qoute("深圳")
data="province="+province+"&city="+city
data=data.encode()
urllib.request.urlopen("http://127.0.0.1:5000",data=data)
```

这里，data=data.encode()是把字符串 data 按 UTF-8 编码转换为二进制数据。POST 方法与 GET 方法最大的不同是：GET 方法的参数放在地址栏的后面，而 POST 的数据放在 urlopen()函数的参数 data 中，而且这个参数值必须是二进制数据。

编程客户端 client.py 程序如下：

```
import urllib.parse
import urllib.request
url="http://127.0.0.1:5000"
try:
    province= urllib.parse.quote("广东")
    city= urllib.parse.quote("深圳")
    data="province="+province+"&city="+city
    data=data.encode()
    html=urllib.request.urlopen("http://127.0.0.1:5000",data=data)
    html = html.read()
    html = html.decode()
    print(html)
except Exception as err:
```

```
            print(err)
```

1.4.2 服务器获取 POST 方法的数据

服务器用 Flask 库中 request 对象的 form 来存储 POST 方法的参数，用 GET 方法来获取参数，即用 flask.request.form.get（参数）来获取参数的值，例如：

```
province=flask.request.form.get("province")
city=flask.request.form.get("city")
```

可以获取 POST 方法传递的参数 province 与 city 的值。

编写服务器程序 server.py 如下：

```
import flask
app=flask.Flask(__name__)

@app.route("/",methods=["POST"])
def index():
    try:
        province=flask.request.form.get("province") if "province" in flask.request.form else ""
        city = flask.request.form.get("city") if "city" in flask.request.form else ""
        return province+","+city
    except Exception as err:
        return str(err)

if __name__=="__main__":
    app.run()
```

注意，在服务器中要指定：

```
@app.route("/",methods=["POST"])
def index():
```

表明这个函数接收 POST 请求。默认时只接收 GET 请求，如果要接收 POST 请求就必须明确指明，如果写成：

```
@app.route("/",methods=["GET","POST"])
def index():
```

那么这个函数既可以接收 GET 请求，也可以接收 POST 请求。

先运行 server.py 建立 Web 网站，再运行 client.py，可以看到 client.py 程序运行结果：

```
广东,深圳
```

微课 6
GET 与 POST 的混合使用

1.4.3 GET 与 POST 方法的混合使用

实际应用中，客户端会同时使用 GET 与 POST 方法向服务器发送数据。一般使用

GET 方法发送的数据，参数简单，数据量少。而使用 POST 方法发送的数据参数复杂，数据量大。

例如，客户端向服务器发送一个城市的简介，服务器接收后返回收到的信息。

我们把省份与城市的名称放在 URL 地址后面，使用 GET 方法发送给服务器，然后把城市的简介用 POST 方法发送给服务器，客户端程序如下：

```
import urllib.parse
import urllib.request
url="http://127.0.0.1:5000"
note="深圳依山傍海，气候宜人，实在是适合人类居住的绝佳地。这里四季如春，干净整洁，比邻香港，拥有着丰富的自然景观和人文气息。匆匆过客注意到的也许只有它的时尚繁华，忙碌的暂居者也可能对它有着不识城市真面目之感。只有世代在此生活的老深圳人，才默默地看着它从贫穷走向富饶经历了怎样的艰辛。"
try:
    province= urllib.parse.quote("广东")
    city= urllib.parse.quote("深圳")
    note= "note="+urllib.parse.quote(note)
    param="province="+province+"&city="+city
    html=urllib.request.urlopen("http://127.0.0.1:5000?"+param,data=note.encode())
    html = html.read()
    html = html.decode()
    print(html)
except Exception as err:
    print(err)
```

服务器程序如下：

```
import flask
app=flask.Flask(__name__)

@app.route("/",methods=["GET","POST"])
def index():
    try:
        province=flask.request.args.get("province") if "province" in flask.request.args else ""
        city = flask.request.args.get("city") if "city" in flask.request.args else ""
        note = flask.request.form.get("note") if "note" in flask.request.form else ""
        return province+","+city+"\n"+note
    except Exception as err:
        return str(err)
```

```
if __name__=="__main__":
    app.run()
```

先运行服务器程序建立 Web 网站,再运行客户端程序,可以看到服务器程序运行结果:

```
广东,深圳
深圳依山傍海,气候宜人,实在是适合人类居住的绝佳地。这里四季如春,干净整洁,比邻香港,拥有着丰富的自然景观和人文气息。匆匆过客注意到的也许只有它的时尚繁华,忙碌的暂居者也可能对它有着不识城市真面目之感。只有世代在此生活的老深圳人,才默默地看着它从贫穷走向富饶经历了怎样的艰辛。
```

显示服务器接收 GET 方法的参数时使用 flask.request.args.get(参数)方法与接收 POST 方法参数的方法 flask.request.form.get(参数)不同。可以把它们统一优化为 flask.request.values.get(参数),使用这个方法可以获取 GET 方法的参数,也可以获取 POST 方法的参数,服务器程序改进如下:

```
import flask
app=flask.Flask(__name__)

@app.route("/",methods=["GET","POST"])
def index():
    try:
        province=flask.request.values.get("province") if "province" in flask.request.values else ""
        city = flask.request.values.get("city") if "city" in flask.request.values else ""
        note = flask.request.values.get("note") if "note" in flask.request.values else ""
        return province+","+city+"\n"+note
    except Exception as err:
        return str(err)

if __name__=="__main__":
    app.run()
```

1.5 搭建图书网站

微课 7
Web 下载服务器程序

PPT Web 下载服务器程序

下面将讲解如何编写爬虫程序去爬取一些实际网站的数据,这样做的好处是让读者有真实的操作体验,缺点是随着时间的推移,这些网站的结构会发生变化,从而导致本书真实案例的爬虫程序不能正常运行。为了避免这种情况的发生,本书编写一个图书网站,展示 50 本书的信息与图片,书中会以该网站为爬取对象,讲解如何使用各种爬虫技术爬取该网站的数据。

1.5.1 准备网站素材

先在 Excel 中准备 50 本书的相关数据，每本书的数据包括编号（ID）、书名（Title）、作者（Author）、出版社（Publisher）、出版日期（PubDate）、价格（Price），如图 1-5-1 所示。把 Excel 文件以 UTF-8 编码格式保存为 books.csv。该文件是文本文件，每个数据用英文逗号分开，每本书的数据占一行。另外准备每本书对应的图片共 50 张，以对应编号命名图片，存储在 static 文件夹中，如图 1-5-2 所示。

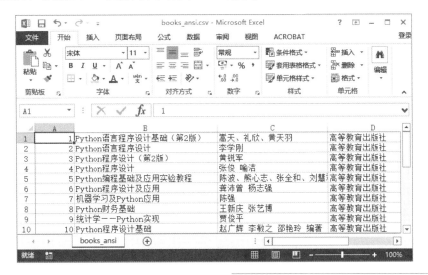

图 1-5-1 图书数据

图 1-5-2 图书图片

1.5.2 创建网页模板

在 Flask 项目文件夹 templates 中创建一个网页模板文件 book.html，这个模板文件包含一组参数 ID、Title、Author、Publisher、PubDate、Price，并使用 books 列表参数构造一个循环，用来显示多本图书。使用 pageIndex 表示当前页码，pageCount 表示总页数，实现翻页功能，文件内容如下：

微课 8
Web 文件客户端程序

```
<style>
```

```
.price { margin: 10px;color:red; }
.attrs { margin: 10px;color: #666; }
.pl {color:#888;}
.link {border: 1px solid ;}
a:link { color: blue; text-decoration: none; }
a:visited { color: blue; text-decoration: none; }
</style>
<div>
<table>
{% for b in books %}
<tr>
    <td><img width="200" src='/static/{{b["Image"]}}'></td>
    <td>
      <div>
            <div class="title"><h3>{{b["Title"]}}</h3></div>
            <div class="author">
                <span class="pl">作者</span>:<span class="attrs">{{b["Author"]}}</span>
            </div>
            <div class="publisher">
                <span class="pl">出版社</span>:<span class="attrs">{{b["Publisher"]}}</span>
            </div>
            <div class="date">
                <span class="pl">出版时间</span>:<span class="attrs">{{b["PubDate"]}}</span>
            </div>
            <div class="price">
                <span class="pl">价格</span>:<span class="price">{{b["Price"]}}</span>
            </div>
      </div>
    </td>
</tr>
{% endfor %}
</table>
</div>
<div align="center" class="paging">
    <a href="/?pageIndex=1" class="link">第一页</a>
    {% if pageIndex>1 %}
        <a href="/?pageIndex={{pageIndex-1}}" class="link">前一页</a>
```

```
        {% else %}
            <a href="#"  class="link">前一页</a>
        {% endif %}
        {% if pageIndex<pageCount %}
            <a href="/?pageIndex={{pageIndex+1}}"  class="link">下一页</a>
        {% else %}
            <a href="#">下一页</a>
        {% endif %}
        <a href="/?pageIndex={{pageCount}}"  class="link">末一页</a>
        <span>Page {{pageIndex}}/{{pageCount}}</span>
</div>
```

1.5.3 创建网站服务器程序

使用 books.csv 文件中的数据，读出全部图书并传递到 book.html 模板，形成多个网页，服务器程序 server.py 如下：

```python
import flask
app=flask.Flask(__name__)

@app.route("/")
def show():
    pageRowCount=3
    pageIndex=int(flask.request.values.get("pageIndex","1"))
    startRow=(pageIndex-1)*pageRowCount
    endRow=pageIndex*pageRowCount
    books=[]
    try:
        fobj=open("books.csv","r",encoding=" UTF-8")
        rows=fobj.readlines()
        count=0
        for row in rows:
            if row.strip("\n").strip()!="":
                count+=1
        count=count-1
        pageCount=count//pageRowCount
        if count % pageRowCount!=0:
            pageCount+=1
        rowIndex=0
        for i in range(1,count+1):
            row=rows[i]
            if rowIndex>=startRow and rowIndex<endRow:
                row=row.strip("\n")
```

微课 9
Web 上传服务器程序 1

微课 10
Web 上传服务器程序 2

```
                            s=row.split(",")
                            m={}
                            m["ID"] = s[0]
                            m["Image"] = s[0] + s[6]
                            m["Title"] = s[1]
                            m["Author"] = s[2]
                            m["PubDate"] = s[3]
                            m["Publisher"] = s[4]
                            m["Price"] = s[5]
                            books.append(m)
                        rowIndex+=1
                    fobj.close()
                except Exception as err:
                    print(err)
                return flask.render_template("book.html",books=books,
            pageIndex=pageIndex,pageCount=pageCount)

            app.debug=True
            app.run()
```

1.5.4 运行网站服务器程序

运行服务器程序可以访问图书网站，如图 1-5-3 所示。单击"下一页"就显示下一页数据，单击"末一页"则显示最后一页的数据。

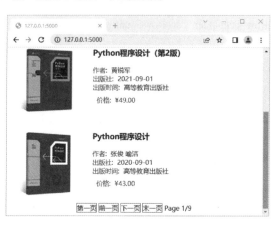

图 1-5-3
图书网站

1.6 正则表达式与查找匹配字符串

1.6.1 正则表达式

正则表达式可以用来匹配与查找指定字符串，从网上爬取数据自然会用到正则表达式。Python 的正则表达式要先导入 re 库，正则表达式以 r 引导，例如：

```
import re
reg=r"\d+"
m=re.search(reg,"abc123cd")
print(m)
```

其中 r"\d+"正则表达式表示匹配连续的多个数值，search()是 re 库中的函数，从 abc123cd 字符串中搜索连续的数值，得到 123，返回一个匹配对象，因此程序的结果如下：

```
<_sre.SRE_Match object; span=(3, 6), match='123'>
```

从结果可以看出，在指定的字符串中找到了连续的数值 123，span(3,6)表示开始位置是 3，结束位置是 6，子字符串是 123 在 abc123cd 中的位置。

Python 中关于正则表达式的规则较多，下面将介绍主要的使用方法，详细使用方法读者可以参考相关资料。

① 使用\d 以匹配 0～9 中的 1 个数值。

例如：

```
import re
reg=r"\d"
m=re.search(reg,"abc123cd")
print(m)
```

匹配结果为字符串中第一个数值 1，程序运行结果：

```
<_sre.SRE_Match object; span=(3, 4), match='1'>
```

② 使用+以重复前 1 个匹配字符一次或多次。

例如：

```
import re
reg=r"b\d+"
m=re.search(reg,"a12b123c")
print(m)
```

匹配结果为字符串中的 b123，程序运行结果：

```
<_sre.SRE_Match object; span=(3, 7), match='b123'>
```

 注意：

r"b\d+"第一个字符要匹配字符 b，后面匹配连续多个数字，因此结果是 b123，而不是 a12。

③ 使用*以重复前 1 个匹配字符零次或多次。

*与+的使用方式类似，例如：

```
import re
reg=r"ab+"
m=re.search(reg,"acabc")
```

17

```
print(m)
reg=r"ab*"
m=re.search(reg,"acabc")
print(m)
```

程序运行结果:

```
<_sre.SRE_Match object; span=(2, 4), match='ab'>
<_sre.SRE_Match object; span=(0, 1), match='a'>
```

可见 r"ab+"匹配结果是 ab,但 r"ab*"匹配结果是 a。因为 r"ab*"表示 b 可以重复零次,但是+却要求 b 重复一次以上。

④ 使用?以重复前 1 个匹配字符零次或一次。

例如:

```
import re
reg=r"ab?"
m=re.search(reg,"abbcabc")
print(m)
```

程序运行结果:

```
<_sre.SRE_Match object; span=(0, 2), match='ab'>
```

程序运行结果是 ab,其中字符 b 重复一次。

⑤ 使用.代表任何 1 个字符,但是没有特别声明时不代表字符\n

例如:

```
import re
s="xaxby"
m=re.search(r"a.b",s)
print(m)
```

可见.代表了字符 x,程序运行结果:

```
<_sre.SRE_Match object; span=(1, 4), match='axb'>
```

⑥ 使用|把字符分成左右两部分。

例如:

```
import re
s="xaababaaby"
m=re.search(r"ab|ba",s)
print(m)
```

程序运行结果:

```
<_sre.SRE_Match object; span=(2, 4), match='ab'>
```

⑦ 特殊字符使用反斜线\引导,例如\r、\n、\t、\\分别表示回车、换行、制表符号与

反斜线。

例如：

```
import re
reg=r"a\nb?"
m=re.search(reg,"ca\nbcabc")
print(m)
```

程序运行结果：

```
<_sre.SRE_Match object; span=(1, 4), match='a\nb'>
```

⑧ 使用\b 表示单词结尾，单词结尾包括各种空白字符或者字符串结尾。

例如：

```
import re
reg=r"car\b"
m=re.search(reg,"The car is black")
print(m)
```

程序运行结果：

```
<_sre.SRE_Match object; span=(4, 7), match='car'>
```

⑨ 使用[]在区间中任选 1 个字符，如果字符是 ASCII 码中连续的一组，那么可以使用-符号连接。例如[0-9]表示 0～9 中的任意 1 个数字，[A-Z]表示 A～Z 的其中 1 个大写字符，[0-9A-Z]表示 0～9 中 1 个数字或 A～Z 的其中 1 个大写字符。

例如：

```
import re
reg=r"x[0-9]y"
m=re.search(reg,"xyx2y")
print(m)
```

程序运行结果：

```
<_sre.SRE_Match object; span=(2, 5), match='x2y'>
```

⑩ ^出现在[]的第一个字符位置，代表取反。例如[^ab0-9]表示不是 a、b，也不是 0～9 中的数字。

例如：

```
import re
reg=r"x[^ab0-9]y"
m=re.search(reg,"xayx2yxcy")
print(m)
```

程序运行结果：

```
<_sre.SRE_Match object; span=(6, 9), match='xcy'>
```

⑪ 使用\s 匹配任何空白字符。

例如：

```
import re
s="1a ba\tbxy"
m=re.search(r"a\sb",s)
print(m)
```

程序运行结果：

<_sre.SRE_Match object; span=(1, 4), match='a b'>

⑫ 使用\w 匹配包括下画线在内的单词字符。

例如：

```
import re
reg=r"\w+"
m=re.search(reg,"Python is easy")
print(m)
```

匹配结果 Python，程序运行结果：

<_sre.SRE_Match object; span=(0, 6), match='Python'>

⑬ 使用^匹配字符串的开头位置。

例如：

```
import re
reg=r"^ab"
m=re.search(reg,"cabcab")
print(m)
```

程序运行结果：

None

没有匹配到任何字符，因为字符串 cabcab 中虽然有 ab，但不是 ab 为开头。

⑭ 使用$匹配字符串的结尾位置。

例如：

```
import re
reg=r"ab$"
m=re.search(reg,"abcab")
print(m)
```

程序运行结果是位于结尾的 ab，而不是开头的 ab，程序运行结果：

<_sre.SRE_Match object; span=(3, 5), match='ab'>

⑮ 使用括号()把()部分进行重复，经常与+、*、?一起使用。

例如：

```
import re
reg=r"(ab)+"
m=re.search(reg,"ababcab")
print(m)
```

程序运行结果 abab，+对 ab 进行了重复，程序运行结果：

```
<_sre.SRE_Match object; span=(0, 4), match='abab'>
```

1.6.2 查找匹配字符串

Python 的正则表达式库 re 的 search()函数使用正则表达式对字符串进行匹配，如果匹配不成功则返回 None，如果匹配成功则返回一个匹配对象。匹配对象调用 start()函数得到匹配字符串的开始位置，匹配对象调用 end()函数得到匹配字符串的结束位置。search()函数虽然只返回第一次匹配的结果，但是连续使用 search()函数可以找到字符串中全部匹配的字符串。

例如，匹配英文句子中所有单词

可以使用正则表达式 r"[A-Za-z]+\b"以匹配单词，该表达式表示匹配由大小写字母组成的连续多个字符，一般是一个单词，\b 表示单词结尾。

```
import re
s="I am testing search function"
reg=r"[A-Za-z]+\b"
m=re.search(reg,s)
while m!=None:
    start=m.start()
    end=m.end()
    print(s[start:end])
    s=s[end:]
    m=re.search(reg,s)
```

程序运行结果：

```
I
am
testing
search
function
```

程序匹配到一个单词，执行 m.start()返回单词起始位置，执行 s[start:end]截取单词，之后程序再次匹配字符串 s[end:]，即字符串的后半段，匹配完毕后可以找出所有单词。

1.7 实践项目——爬取学生信息

微课 11
Web 学生管理程序 1

1.7.1 项目简介

设计 Web 服务器程序 server.py，读取 students.txt 文件中的学生数据，以表格的形式显示在网页上，其中 students.txt 的格式如下：

```
No,Name,Gender,Age
1001,张三,男,20
1002,李四,女,19
1003,王五,男,21
```

第一行是表格的表头，有 No（学号）、Name（姓名）、Gender（性别）、Age（年龄）4 个属性，每个学生数据占一行，各个数据之间用逗号分开。

设计客户端的爬虫程序，从这个网页上爬取学生的信息，并存储到数据库中。学生数据库可以使用 Sqlite 数据库中的 students.db。

1.7.2 服务器程序

服务器程序首先读取同一个目录下的 students.txt 文件，然后组成一张 HTML 表格，用网页的形式呈现，如图 1-7-1 所示。

微课 12
Web 学生管理程序 2

图 1-7-1
学生信息表网页

微课 13
正则表达式

程序先检查是否有 students.txt 文件存在，有则读取，读出的一行的数据是一个字符串用逗号分开，因此可以使用 split(",")函数拆分字符串，然后把一整行数据组织在 <tr>...</tr> 标签中，把每个数据组织在<td>...<td>标签中，程序如下：

```
from flask import Flask,request
import os


app=Flask(__name__)


@app.route("/")
def show():
    if os.path.exists("students.txt"):
        st="<h3>学生信息表</h3>"
```

```python
        st=st+"<table border='1' width='300'>"
        fobj=open("students.txt","rt",encoding="utf-8")
        while True:
            #读取一行,去除行尾部"\n"换行符号
            s=fobj.readline().strip("\n")
            #如果读到文件尾部就退出
            if s=="":
                break
            #按逗号拆分开
            s=s.split(",")
            st=st+"<tr>"
            #把各个数据组织在<td>...</td>标签中
            for i in range(len(s)):
                st=st+"<td>"+s[i]+"</td>"
            #完成一行
            st=st+"</tr>"
        fobj.close()
        st=st+"</table>"
        return st

if __name__=="__main__":
    app.run()
```

微课 14
爬取学生信息 1

源代码 爬取学生信息

PPT 爬取学生信息

运行服务器程序,默认的网址是 http://127.0.0.1:5000/。

1.7.3 客户端程序

客户端程序需要访问 http://127.0.0.1:5000/,获取 HTML 网页,结果如下:

```
<h3>学生信息表</h3><table border='1' width='300'><tr><td>No</td><td>Name</td><td>Gender</td><td>Age</td></tr><tr><td>1001</td><td>张三</td><td>男</td><td>20</td></tr><tr><td>1002</td><td>李四</td><td>女</td><td>19</td></tr><tr><td>1003</td><td>王五</td><td>男</td><td>21</td></tr></table>
```

从该 HTML 网页中爬取数据,先分解出第一行数据如下:

```
<tr><td>No</td><td>Name</td><td>Gender</td><td>Age</td></tr>
```

分解这一行标签<td>...</td>中的数据,可知表头有哪些字段,该表表头字段有 No、Name、Gender、Age。

分解下一行数据如下:

```
<tr><td>1001</td><td>张三</td><td>男</td><td>20</td></tr>
```

再次分解这一行<td>...</td>标签中的数据,得到 No、Name、Gender、Age 对应的数据依次是"1001""张三""男""20",把这一行的数据写入对应的数据库。

微课 15
爬取学生信息 2

分解出标签\<tr>...\</tr>，要使用正则表达式 r"\<tr>"与 r"\</tr>"，先用 r"\<tr>"匹配 HTML 代码，得到第一个标签\<tr>的位置，再使用 r"\</tr>"匹配 HTML 字符串，得到第一个标签\</tr>的位置，取出标签\<tr>...\</tr>中的数据。再次使用正则表达式 r"\<td>"与 r"\</td>"分解标签\<td>...\</td>中的数据。

客户端程序如下：

```python
import urllib.request
import re
import sqlite3

def searchWeb(html):
    rows=[]
    #查询第一个<tr>...</tr>行
    m=re.search(r"<tr>",html)
    n=re.search(r"</tr>",html)
    if m!=None and n!=None:
        #跳过第一行的标题
        html=html[n.end():]
        # 查询第二行开始的数据部分
        m=re.search(r"<tr>",html)
        n=re.search(r"</tr>",html)
        while(m!=None and n!=None):
            row=[]
            #start 是<tr>的结束位置
            start=m.end()
            #end 是</tr>的开始位置
            end=n.start()
            #t 是<tr>...</tr>包含的字符串
            t=html[start:end]
            #html[n.end():]是剩余的 html
            html=html[n.end():]
            #查询第一组<td>...</td>
            a=re.search(r"<td>",t)
            b=re.search(r"</td>",t)
            i=0
            while (a!=None and b!=None):
                start=a.end()
                end=b.start()
                #找到一组<td>...</td>的数据
                row.append(t[start:end])
                #t[b.end():]是本行剩余的部分
```

```
                t = t[b.end():]
                a = re.search(r"<td>", t)
                b = re.search(r"</td>", t)
        #增加一行数据
        rows.append(row)
        #继续查找下一行<tr>...</tr>
        m = re.search(r"<tr>", html)
        n = re.search(r"</tr>", html)
    return rows

def saveDB(rows):
    if len(rows)==0:
        #没有数据就返回
        return
    try:
        con = sqlite3.connect("students.db")
        cursor = con.cursor()
        try:
            #如果有 students 表就删除
            cursor.execute("drop table students")
        except:
            pass
        try:
            #建立新的 students 表
            sql = "create table students (No varchar(128) primary key,Name varchar(128),Gender varchar(128),Age int)"
            cursor.execute(sql)
        except:
            pass

        for row in rows:
            if(len(row)==4):
                #插入一条记录
                sql="insert into students (No,Name,Gender,Age) values (?,?,?,?)"
                try:
                    No=row[0]
                    Name=row[1]
                    Gender=row[2]
                    Age=int(row[3])
                    cursor.execute(sql,(No,Name,Gender,Age))
                except Exception as err:
```

```python
                    print(err)
                #数据库提交保存
                con.commit()
                con.close()
        except Exception as err:
            print(err);


def showWeb(rows):
    print("Showing data from Web...")
    for row in rows:
        print(row)


def showDB():
    print("Showing data from DB...")
    try:
        con = sqlite3.connect("students.db")
        cursor = con.cursor()
        #查询数据库记录
        cursor.execute("select * from students")
        rows=cursor.fetchall()
        #显示每条记录
        for row in rows:
            print(row)
        con.close()
    except Exception as err:
        print(err)


try:
    url = "http://127.0.0.1:5000"
    #访问这个网址获取 html
    resp=urllib.request.urlopen(url)
    data=resp.read()
    html=data.decode("utf-8")
    #在 html 中查找学生信息
    rows=searchWeb(html)
    #显示查找的信息
    showWeb(rows)
    #保存学生信息到数据库
    saveDB(rows)
    #显示数据库的数据
    showDB()
```

```
except Exception as e:
    print(e)
```

客户端从服务器的网页中爬取了学生信息并保存到数据库，程序运行结果：

```
Showing data from Web...
['1001', '张三', '男', '20']
['1002', '李四', '女', '19']
['1003', '王五', '男', '21']
Showing data from DB...
('1001', '张三', '男', 20)
('1002', '李四', '女', 19)
('1003', '王五', '男', 21)
```

注意：

通过使用正则表达式匹配的方法来爬取网页的数据会比较麻烦，在后面的章节中将介绍更加简单高效的爬取方法。

练习一

① Flask 是 Python 一个比较简单的 Web 程序开发框架，请列出其他 Python 流行的 Web 开发框架，并说明主要区别。

② 使用 GET 与 POST 方法提交数据有什么不同？

③ 使用 Flask 编写一个 Web 程序，接收如下表单提交的 user_name 与 user_pass 数据：

```
<form name="frm" method="post" action="/login">
<input type="text" name="user_name" >
<input type="password" name="user_pass">
<input type="submit" value="Login" >
</form>
```

④ 说明如下正则表达式匹配的字符串是什么？

- r"\w+\s"
- r"\w+\b"
- r"\d+-\d+"
- r"\w+@(\w+\.)+\w+"
- r"(b|cd)ef"

⑤ 使用正则表达式匹配 HTML 网页中形如<img src="http://.../文件名.jpg">的图片文件，找出图片文件的网址，并下载这些图片。

项目 2　爬取天气预报数据

本项目介绍如何使用 BeautifulSoup 库解析 HTML 文档与爬取文档数据的方法，最后介绍爬取中国天气网的城市天气预报数据的综合案例。

2.1 HTML 文档结构与文档树

2.1.1 HTML 文档结构

微课 16
HTML 文档结构

HTML 文档实际上类似 XML 文档，完整的 HTML 文档包含根元素<html>，在<html>中包含<head>、<body>等标签，一个典型的 HTML 文档如下所示：

```
<html><head><title>The Dormouse's story</title></head>
<body>
<p class="title"><b>The Dormouse's story</b></p>
<p class="story">
Once upon a time there were three little sisters; and their names were
<a href="http://example.com/elsie" class="sister" id="link1">Elsie</a>,
<a href="http://example.com/lacie" class="sister" id="link2">Lacie</a> and
<a href="http://example.com/tillie" class="sister" id="link3">Tillie</a>;
and they lived at the bottom of a well.
</p>
<p class="story">...</p>
</body>
</html>
```

该文档在浏览器中显示的效果，如图 2-1-1 所示。

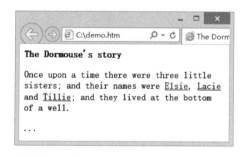

图 2-1-1
HTML 文档

HTML 文档中<...>元素称为 tag 元素或者 element 元素，例如<html>、<body>、<title>、<p>、<a>等都是 tag 元素，每个 tag 元素都有对应的结束元素</...>，例如</html>、</body>、</title>、</p>、</a>等。

一个 tag 元素可以有很多属性，例如<p class="title">中的<p>元素有属性 class，属性值为 title。

📝 注意：

HTML 中除了 tag 元素，穿插于 tag 元素之间的文本也是元素，称为 text 元素，例如<title>The Dormouse's story</title>中的文本 The Dormouse's story 是一个 text 文本元素，其父节点是<title>。

HTML 中的 tag 元素的名称是不区分大小写的，因此<html>、<HTML>、<Html>是一样的，这一点与 XML 不同。

2.1.2 HTML 文档树

HTML 的结构是一个树结构，例如：

```
<html>
<head><title>Demo</title></head>
<body>
<div>A<p>B</p>C</div>
<span>D</span>
</body>
</html>
```

那么对应的文档树如图 2-1-2 所示。

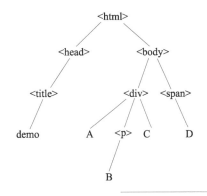

图 2-1-2
HTML 文档树

HTML 文档树的概念是十分重要的，是查找 tag 节点的重要依据。

2.2 BeautifulSoup 装载 HTML 文档

HTML 文档节点的查找工具很多，BeautifulSoup 是功能强大且十分流行的查找工具。本节介绍 BeautifulSoup 的基本应用。

2.2.1 BeautifulSoup 的安装

BeautifulSoup 是 Python 的第三方工具，在 bs4 库中，需要安装 bs4 库。安装流程如下，先进入 Python 的安装目录（例如，Python 在 C:\Python36），再进入 Scripts 子目录，找到 pip 程序，执行：

```
pip install bs4
```

成功安装后就可以在 Python 的命令行中测试语句：

```
from bs4 import BeautifulSoup
```

如果执行这条语句没有报错，说明安装 bs4 库成功。

2.2.2 BeautifulSoup 装载 HTML 文档

doc 是一个 HTML 文档，通过：

微课 17
BeautifulSoup 的安装

PPT　BeautifulSoup
的安装

```
from bs4 import BeautifulSoup
soup=BequtifulSoup(doc,"lxml")
```

就可以创建一个名称为 soup 的 BeautifulSoup 对象,其中 doc 是一个 HTML 文档字符串,lxml 是一个参数,表示创建的是一个通过 lxml 解析器解析的文档。BeautifulSoup 有多种解析器,其中 lxml 是最常用的一个。

通过调用 soup.prettify()可以把 soup 对象的文档树变成一个字符串。

微课 18
BeautifulSoup 装载
HTML 文档 1

例 2-2-1: 用 BeautifulSoup 装载 HTML 文档,显示文档的树状结构

```
from bs4 import BeautifulSoup
doc='''
<html><head><title>The Dormouse's story</title></head>
<body>
<p class="title"><b>The Dormouse's story</b></p>
<p class="story">
Once upon a time there were three little sisters; and their names were
<a href="http://example.com/elsie" class="sister" id="link1">Elsie</a>,
<a href="http://example.com/lacie" class="sister" id="link2">Lacie</a> and
<a href="http://example.com/tillie" class="sister" id="link3">Tillie</a>;
and they lived at the bottom of a well.
</p>
<p class="story">...</p>
</body>
</html>
'''
soup=BeautifulSoup(doc,"lxml")
s=soup.prettify()
print(s)
```

程序运行结果:

```
<html>
 <head>
  <title>
   The Dormouse's story
  </title>
 </head>
 <body>
  <p class="title">
   <b>
    The Dormouse's story
```

```
    </b>
   </p>
   <p class="story">
    Once upon a time there were three little sisters; and their names were
    <a class="sister" href="http://example.com/elsie" id="link1">
     Elsie
    </a>
    ,
    <a class="sister" href="http://example.com/lacie" id="link2">
     Lacie
    </a>
    and
    <a class="sister" href="http://example.com/tillie" id="link3">
     Tillie
    </a>
    ;
and they lived at the bottom of a well.
   </p>
   <p class="story">
    ...
   </p>
  </body>
 </html>
```

可见 BeautifulSoup 装载了 HTML 文档,最后通过 prettify() 函数把文档树转为字符串的格式。

BeautifulSoup 装载文档的功能十分强大,在装载的过程中如果发现 HTML 文档中的元素有缺失的情况,会尽可能地对文档进行修复,使得最后的文档树是一棵完整的树。这一点十分重要,因为面临的大多数网页都或多或少有元素缺失的情况,但 BeautifulSoup 都能正确装载它们。

微课 19
BeautifulSoup 装载
HTML 文档 2

例 2-2-2: BeautifulSoup 装载有元素缺失的 HTML 文档

```
from bs4 import BeautifulSoup
doc='''
<title>有缺失元素的 HTML 文档</title>
<div>
<A href='one.html'>one</a>
<p>
<a href='two.html'>two</a>
</DIV>
'''
soup=BeautifulSoup(doc,"lxml")
```

```
s=soup.prettify()
print(s)
```

程序运行结果：

```
html>
 <head>
  <title>
   有缺失元素的 HTML 文档
  </title>
 </head>
 <body>
  <div>
   <a href="one.html">
    one
   </a>
   <p>
    <a href="two.html">
     two
    </a>
   </p>
  </div>
 </body>
</html>
```

可见在 HTML 缺失<html>根元素时，BeautifulSoup 会自动补全。在发现 HTML 中<title>元素没有<head>元素后自动补全，用<head>元素包含<title>元素。在发现 HTML 中没有<body>元素时，BeautifulSoup 也自动补全，还将<A href='one.html'>one</a>优化成<a href='one.html'>one</a>，同时优化如下代码：

```
<p>
<a href='two.html'>two</a>
```

将该代码改写成：

```
    <p>
     <a href="two.html">
      two
     </a>
    </p>
```

BeautifulSoup 还将最后的</DIV>改成</div>，默认情况下将所有的 tag 元素的名称都改为小写。通过 BeautifulSoup 的修正后这棵 HTML 树就比较完整了。

值得注意的是 BeautifulSoup 虽然功能强大，能修正一些缺失的 HTML 元素，但是还没有智能到完全修复所有 HTML 文档错误的程度。

2.3 查找文档元素

2.3.1 查找 HTML 元素

查找 HTML 元素是爬取网页信息的重要手段，BeautifulSoup 提供了一系列查找 HTML 元素的方法，其中功能强大的 find_all()函数就是其中常用的一个方法。find_all()函数的使用方法如下：

```
find_all(self, name=None, attrs={}, recursive=True, text=None, limit=None, **kwargs)
```

- self 表明该函数是一个类成员函数。
- name 是要查找的 HTML 元素名称，默认值为 None，如果不提供，就查找所有 HTML 元素。
- attrs 是元素的属性，数据结构是一个字典，默认值为空，如果提供就查找有指定属性的 HTML 元素。
- recursive 在元素节点的子树下面是否全范围进行查找，默认值为 True。

find_all()函数返回查找到所有指定的元素列表，列表中每个元素的数据结构是 bs4.element.Tag 对象。

find_all()函数功能是查找所有满足要求的 HTML 元素节点，如果只需要查找一个 HTML 元素节点，可以使用 BuautifulSoup 中的 find()函数、find()函数使用方法如下：

```
find(self, name=None, attrs={}, recursive=True, text=None, limit=None, **kwargs)
```

find()函数中参数的数据结构使用方法与 find_all()类似，不同的是 find()只返回第一个满足要求的元素节点，而不是一个列表。

例 2-3-1： 查找文档中的<title>元素

```
from bs4 import BeautifulSoup
doc='''
<html><head><title>The Dormouse's story</title></head>
<body>
<p class="title"><b>The Dormouse's story</b></p>
<p class="story">
Once upon a time there were three little sisters; and their names were
<a href="http://example.com/elsie" class="sister" id="link1">Elsie</a>,
<a href="http://example.com/lacie" class="sister" id="link2">Lacie</a> and
<a href="http://example.com/tillie" class="sister" id="link3">Tillie</a>;
and they lived at the bottom of a well.
</p>
<p class="story">...</p>
</body>
</html>
'''
```

```
soup=BeautifulSoup(doc,"lxml")
tag=soup.find("title")
print(type(tag),tag)
```

程序运行结果：

```
<class 'bs4.element.Tag'> <title>The Dormouse's story</title>
```

可见查找到<title>元素，返回元素数据结构是一个 bs4.element.Tag 对象。

例 2-3-2： 查找文档中的所有<a>元素

```
from bs4 import BeautifulSoup
doc='''
<html><head><title>The Dormouse's story</title></head>
<body>
<p class="title"><b>The Dormouse's story</b></p>
<p class="story">
Once upon a time there were three little sisters; and their names were
<a href="http://example.com/elsie" class="sister" id="link1">Elsie</a>,
<a href="http://example.com/lacie" class="sister" id="link2">Lacie</a> and
<a href="http://example.com/tillie" class="sister" id="link3">Tillie</a>;
and they lived at the bottom of a well.
</p>
<p class="story">...</p>
</body>
</html>
'''
soup=BeautifulSoup(doc,"lxml")
tags=soup.find_all("a")
for tag in tags:
    print(tag)
```

程序运行结果是查找到 3 个<a>元素：

```
<a class="sister" href="http://example.com/elsie" id="link1">Elsie</a>
<a class="sister" href="http://example.com/lacie" id="link2">Lacie</a>
<a class="sister" href="http://example.com/tillie" id="link3">Tillie</a>
```

例 2-3-3： 查找文档中的第一个<a>元素

```
from bs4 import BeautifulSoup
doc='''
<html><head><title>The Dormouse's story</title></head>
<body>
<p class="title"><b>The Dormouse's story</b></p>
<p class="story">
```

```
Once upon a time there were three little sisters; and their names were
<a href="http://example.com/elsie" class="sister" id="link1">Elsie</a>,
<a href="http://example.com/lacie" class="sister" id="link2">Lacie</a> and
<a href="http://example.com/tillie" class="sister" id="link3">Tillie</a>;
and they lived at the bottom of a well.
</p>
<p class="story">...</p>
</body>
</html>
'''
soup=BeautifulSoup(doc,"lxml")
tag=soup.find("a")
print(tag)
```

找到第一个<a>元素,程序运行结果是:

```
<a class="sister" href="http://example.com/elsie" id="link1">Elsie</a>
```

例 2-3-4: 查找文档中 class="title"的<p>元素

```
from bs4 import BeautifulSoup
doc='''
<html><head><title>The Dormouse's story</title></head>
<body>
<p class="title"><b>The Dormouse's story</b></p>
<p class="story">
Once upon a time there were three little sisters; and their names were
<a href="http://example.com/elsie" class="sister" id="link1">Elsie</a>,
<a href="http://example.com/lacie" class="sister" id="link2">Lacie</a> and
<a href="http://example.com/tillie" class="sister" id="link3">Tillie</a>;
and they lived at the bottom of a well.
</p>
<p class="story">...</p>
</body>
</html>
'''
soup=BeautifulSoup(doc,"lxml")
tag=soup.find("p",attrs={"class":"title"})
print(tag)
```

找到 class="title"的<p>元素,程序运行结果是

```
<p class="title"><b>The Dormouse's story</b></p>
```

如果使用:

```
tag=soup.find("p")
```

也能找到该元素,因为该元素是文档中第一个<p>元素。

例 2-3-5: 查找文档中的 class="sister"元素

```
from bs4 import BeautifulSoup
doc='''
<html><head><title>The Dormouse's story</title></head>
<body>
<p class="title"><b>The Dormouse's story</b></p>
<p class="story">
Once upon a time there were three little sisters; and their names were
<a href="http://example.com/elsie" class="sister" id="link1">Elsie</a>,
<a href="http://example.com/lacie" class="sister" id="link2">Lacie</a> and
<a href="http://example.com/tillie" class="sister" id="link3">Tillie</a>;
and they lived at the bottom of a well.
</p>
<p class="story">...</p>
</body>
</html>
'''
soup=BeautifulSoup(doc,"lxml")
tags=soup.find_all(name=None,attrs={"class":"sister"})
for tag in tags:
    print(tag)
```

在使用 find_all()函数时,参数 name=None 表示无论什么名字的元素都需要查找到,查找到的元素有 3 个:

```
<a class="sister" href="http://example.com/elsie" id="link1">Elsie</a>
<a class="sister" href="http://example.com/lacie" id="link2">Lacie</a>
<a class="sister" href="http://example.com/tillie" id="link3">Tillie</a>
```

在该文档查找 class="sister"元素,使用语句:

```
tags=soup.find_all("a")
```

或:

```
tags=soup.find_all("a",attrs={"class":"sister"})
```

返回结果相同。

2.3.2 获取 HTML 元素属性值

在找到一个元素后,可以在 BeautifulSoup 中获取该元素的属性值。使用方法如下:

```
tag[attrName]
```

获取 tag 元素中名为 attrName 的属性值,其中 tag 是一个数据结构为 bs4.element.Tag 的对象。

例 2-3-6:查找文档中所有超链接地址

```
from bs4 import BeautifulSoup
doc='''
<html><head><title>The Dormouse's story</title></head>
<body>
<p class="title"><b>The Dormouse's story</b></p>
<p class="story">
Once upon a time there were three little sisters; and their names were
<a href="http://example.com/elsie" class="sister" id="link1">Elsie</a>,
<a href="http://example.com/lacie" class="sister" id="link2">Lacie</a> and
<a href="http://example.com/tillie" class="sister" id="link3">Tillie</a>;
and they lived at the bottom of a well.
</p>
<p class="story">...</p>
</body>
</html>
'''
soup=BeautifulSoup(doc,"lxml")
tags=soup.find_all("a")
for tag in tags:
    print(tag["href"])
```

程序运行结果:

```
http://example.com/elsie
http://example.com/lacie
http://example.com/tillie
```

2.3.3 获取元素包含的文本值

一个元素已查找到,例如已查找到<a>元素,那么,怎样获取该元素包含的文本值呢?在 BeautifulSoup 中使用方法如下:

```
tag.text
```

微课 20
BeautifulSoup 获取元素包含的文本值 1

获取 tag 元素包含的文本值, tag 是一个数据结构为 bs4.element.Tag 的对象。

例 2-3-7:查找文档中所有<a>元素中包含的文本值

```
from bs4 import BeautifulSoup
doc='''
<html><head><title>The Dormouse's story</title></head>
<body>
```

微课 21
BeautifulSoup 获取元素包含的文本值 2

```
<p class="title"><b>The Dormouse's story</b></p>
<p class="story">
Once upon a time there were three little sisters; and their names were
<a href="http://example.com/elsie" class="sister" id="link1">Elsie</a>,
<a href="http://example.com/lacie" class="sister" id="link2">Lacie</a> and
<a href="http://example.com/tillie" class="sister" id="link3">Tillie</a>;
and they lived at the bottom of a well.
</p>
<p class="story">...</p>
</body>
</html>
'''
soup=BeautifulSoup(doc,"lxml")
tags=soup.find_all("a")
for tag in tags:
    print(tag.text)
```

程序运行结果：

```
Elsie
Lacie
Tillie
```

如果 tag 元素包含的不是一个简单的文本字符串，而是复杂的结构，那么在使用 tag.text 后返回的是 tag 子树下所有文本节点组合的字符串。

例 2-3-8：查找文档中所有<p>元素中包含的文本值

```
from bs4 import BeautifulSoup
doc='''
<html><head><title>The Dormouse's story</title></head>
<body>
<p class="title"><b>The Dormouse's story</b></p>
<p class="story">
Once upon a time there were three little sisters; and their names were
<a href="http://example.com/elsie" class="sister" id="link1">Elsie</a>,
<a href="http://example.com/lacie" class="sister" id="link2">Lacie</a> and
<a href="http://example.com/tillie" class="sister" id="link3">Tillie</a>;
and they lived at the bottom of a well.
</p>
<p class="story">...</p>
</body>
</html>
'''
```

```
soup=BeautifulSoup(doc,"lxml")
tags=soup.find("p")
for tag in tags:
    print(tag.text)
```

程序运行结果：

```
The Dormouse's story

Once upon a time there were three little sisters; and their names were
Elsie,
Lacie and
Tillie;
and they lived at the bottom of a well.


...
```

列表 tags 中第二个值包含的就是第二个<p>节点子树下所有文本节点的组合值。

2.3.4 高级查找

find()或者 find_all()函数能满足一般需要，如果这两个函数不能满足需求，那么可以设计一个查找函数来完成任务。

例 2-3-9： 查找文档中属性值为 href="http://example.com/lacie"的节点元素<a>

微课 22
BeautifulSoup 高级
查找 1

```
from bs4 import BeautifulSoup
doc='''
<html><head><title>The Dormouse's story</title></head>
<body>
<a href="http://example.com/elsie" >Elsie</a>
<a href="http://example.com/lacie" >Lacie</a>
<a href="http://example.com/tillie" >Tillie</a>
</body>
</html>
'''
def myFilter(tag):
    print(tag.name)
    return (tag.name=="a" and tag.has_attr("href") and tag["href"]=="http://example.com/lacie")

soup=BeautifulSoup(doc,"lxml")
tag=soup.find_all(myFilter)
print(tag)
```

程序运行结果：

```
html
head
title
body
a
a
a
[<a href="http://example.com/lacie">Lacie</a>]
```

程序中定义了筛选函数 myFilter(tag)，该函数的参数是 tag 对象，在调用 soup.find_all(myFilter)时程序会把每个 tag 元素传递给 myFilter()函数，由该函数决定这个 tag 的取舍，如果 myFilter()返回 True 就保留这个 tag 到结果列表中，不然就丢掉这个 tag。因此程序执行时可以看到 html、body、head、title、body、a 等 tag 经过 myFilter()的筛选，只有节点<a href="http://example.com/lacie">Lacie</a>满足要求，因此返回结果为：

```
[<a href="http://example.com/lacie">Lacie</a>]
```

- tag.name 是 tag 的名称。
- tag.has_attr(attName)判断 tag 是否有 attName 属性。
- tag[attName]是 tag 的 attName 属性值。

例 2-3-10：查找文本值以 cie 结尾所有<a>节点

微课 23
BeautifulSoup 高级
查找 2

```python
from bs4 import BeautifulSoup
doc='''
<html><head><title>The Dormouse's story</title></head>
<body>
<a href="http://example.com/elsie" >Elsie</a>
<a href="http://example.com/lacie" >Lacie</a>
<a href="http://example.com/tillie" >Tillie</a>
<a href="http://example.com/tilcie" >Tilcie</a>
</body>
</html>
'''
def endsWith(s,t):
    if len(s)>=len(t):
        return s[len(s)-len(t):]==t
    return False


def myFilter(tag):
    return (tag.name=="a" and endsWith(tag.text,"cie"))


soup=BeautifulSoup(doc,"lxml")
tags=soup.find_all(myFilter)
```

```
for tag in tags:
    print(tag)
```

程序运行结果：

```
<a href="http://example.com/lacie">Lacie</a>
<a href="http://example.com/tilcie">Tilcie</a>
```

程序中定义了一个 endsWith(s,t)函数，判断 s 字符串是否以字符串 t 结尾，若是则返回 True，不是则返回 False。在 myFilter()中调用该函数判断 tag.text 中的文本是否以 cie 结尾，最后找出所有文本值以 cie 结尾的<a>节点。

2.4 遍历文档元素

2.4.1 获取元素节点的父节点元素

在 BeautifulSoup 中使用如下方法：

```
tag.parent
```

获取 tag 节点的父节点元素，根节点<html>的父节点是名称为[document]的节点，这个[document]节点的父节点是 None。

例 2-4-1： 查找文档中<p class="title"><b>The Dormouse's story</b></p>的<b>元素节点的所有父节点的名称。

PPT　获取元素节点的父节点元素

微课 24 获取元素节点的父节点 1

```
from bs4 import BeautifulSoup
doc='''
<html><head><title>The Dormouse's story</title></head>
<body>
<p class="title"><b>The Dormouse's story</b></p>
<p class="story">
Once upon a time there were three little sisters; and their names were
<a href="http://example.com/elsie" class="sister" id="link1">Elsie</a>,
<a href="http://example.com/lacie" class="sister" id="link2">Lacie</a> and
<a href="http://example.com/tillie" class="sister" id="link3">Tillie</a>;
and they lived at the bottom of a well.
</p>
<p class="story">...</p>
</body>
</html>
'''
soup=BeautifulSoup(doc,"lxml")
print(soup.name)
tag=soup.find("b")
```

```
        while tag:
            print(tag.name)
            tag=tag.parent
```

程序运行结果：

```
[document]
b
p
body
html
[document]
```

可见<b>节点的父节点依次为<p>、<body>、<html>。

2.4.2 获取元素节点的所有子节点元素

在 BeantifulSoup 中使用如下方法：

```
tag.children
```

获取 tag 节点的所有子节点元素，包括 element、text 等类型的节点。

例 2-4-2： 获取<p>元素所有直接子节点元素

```
from bs4 import BeautifulSoup
doc='''
<html><head><title>The Dormouse's story</title></head>
<body>
<p class="title"><b>The <i>Dormouse's</i> story</b> Once upon a time ...</p>
</body>
</html>
'''
soup=BeautifulSoup(doc,"lxml")
tag=soup.find("p")
for x in tag.children:
    print(x)
```

程序运行结果：

```
<b>The <i>Dormouse's</i> story</b>
Once upon a time ...
```

<p>节点下面有两个子节点元素，一个是 element 类型的节点<b>The <i>Dormouse's</i> story</b>，另一个是 text 类型的节点 Once upon a time...。

2.4.3 获取元素节点的所有子孙节点元素

在 BeautifulSoup 中使用如下方法：

> tag.desendants

获取 tag 节点的所有子孙节点元素，包括 element、text 等类型的节点。

例 2-4-3：获取<p>元素中的所有子孙节点元素

```
from bs4 import BeautifulSoup
doc='''
<html><head><title>The Dormouse's story</title></head>
<body>
<p class="title"><b>The <i>Dormouse's</i> story</b> Once upon a time ...</p>
</body>
</html>
'''
soup=BeautifulSoup(doc,"lxml")
tag=soup.find("p")
for x in tag.descendants:
    print(x)
```

程序运行结果：

```
<b>The <i>Dormouse's</i> story</b>
The
<i>Dormouse's</i>
Dormouse's
story
Once upon a time ...
```

可见<p>节点元素有如下几个子孙节点：

- The：text 孙节点，是<b>的子节点。
- <i>Dormouse's</i>：element 孙节点，是<b>的子节点。
- Dormouse's：text 孙节点，<i>Dormouse's</i>的子节点。
- story：text 孙节点，是<b>的子节点。
- Once upon a time ...：text 子节点。

2.4.4 获取元素节点的兄弟节点

在 BeautifulSoup 中使用如下方法：

> tag.next_sibling
> tag.previous_sibling

分别是获取下一个和上一个兄弟节点，其中 tag.next_sibling 是 tag 临近的下一个兄弟节点，tag.previous_sibling 是 tag 的临近的上一个兄弟节点。

例 2-4-4：查找前后兄弟节点

```
from bs4 import BeautifulSoup
doc='''
```

```
<html><head><title>The Dormouse's story</title></head>
<body>
<p class="title"><b>The <i>Dormouse's</i> story</b> Once upon a time ...</p>
</body>
</html>
'''
soup=BeautifulSoup(doc,"lxml")
tag=soup.find("b")
print(tag.previous_sibling)
print(tag.next_sibling)
tag=soup.find("i")
print(tag.previous_sibling)
print(tag.next_sibling)
```

程序运行结果：

```
None
Once upon a time ...
The
story
```

可见<b>节点的上一个兄弟节点为 None，下一个兄弟节点为 text 节点 Once upon a time...，<i>节点的上一个兄弟节点是 text 节点 The，下一个是 text 节点 story。

2.5 使用 CSS 语法查找元素

2.5.1 使用 CSS 语法

微课 26
使用 CSS 语法 1

BeautifulSoup 除了可以用 find()与 find_all()函数查找 HTML 文档树中节点元素，还可以使用 CSS 类似的语法来查询，使用方法如下：

```
tag.select(css)
```

其中 tag 是一个 bs4.element.Tag 对象，即 HTML 中的一个 element 节点元素，select 是查找方法，css 是类似 CSS 语法的一个字符串，结构如下：

```
[tagName][attName[=value]]
```

- [...]中的部分是可选的。
- tagName 是元素名称，如果没有指定就是查找所有元素。
- attName=value 是属性名称，value 是对应的值，可以不指定属性，也可以不指定值。
- tag.select(css)返回一个 bs4.element.Tag 对象的列表。

例 2-5-1：使用 soup.select("a") 查找文档中所有<a>元素节点
- soup.select("p a") 查找文档中所有<p>节点下的所有<a>节点元素。

2.5 使用 CSS 语法查找元素

- soup.select("p[class='story'] a") 查找文档中所有属性为 class="story" 的<p>节点下的所有<a>节点元素。
- soup.select("p[class] a") 查找文档中所有具有 class 属性的<p>节点下的所有<a>节点元素。
- soup.select("a[id='link1']") 查找属性 id="link1" 的<a>节点。
- soup.select("body head title") 查找<body>的 <head>中的<title>节点。
- soup.select("body [class] ") 查找<body>的所有具有 class 属性的节点。
- soup.select("body [class] a") 查找<body>的所有具有 class 属性的节点下的<a>节点。

微课 27
使用 CSS 语法 2

例 2-5-2： 查找 HTML 文档中所有<p>中的<a>的链接

```
from bs4 import BeautifulSoup
doc='''
<html><head><title>The Dormouse's story</title></head>
<body>
<p class="title"><b>The Dormouse's story</b></p>
<p class="story">
Once upon a time there were three little sisters; and their names were
<a href="http://example.com/elsie" class="sister" id="link1">Elsie</a>,
<a href="http://example.com/lacie" class="sister" id="link2">Lacie</a> and
<a href="http://example.com/tillie" class="sister" id="link3">Tillie</a>;
and they lived at the bottom of a well.
</p>
<p class="story">...</p>
</body>
</html>
'''
soup=BeautifulSoup(doc,"lxml")
tags=soup.select("p[class='story'] a")
for tag in tags:
    print(tag["href"])
```

程序运行结果：

```
http://example.com/elsie
http://example.com/lacie
http://example.com/tillie
```

也可通过使用以下方法得到一样的结果：

```
tags=soup.select("p a")
tags=soup.select("a")
tags=soup.select("p[class] a")
```

2.5.2 属性的语法规则

在 CSS 结构中的[attName=value]中表示属性 attrName 与 value 相等,也可以指定不相等、包含等运算关系,使用方法如表 2-5-1 所示:

表 2-5-1 CSS 结构使用方法

选择器	功能
[attName]	用于选取带有指定属性的元素
[attName=value]	用于选取带有指定属性和值的元素
[attName^=value]	匹配属性值以指定值开头的每个元素
[attName$=value]	匹配属性值以指定值结尾的每个元素
[attrName*=value]	匹配属性值中包含指定值的每个元素

将该语法套入到 soup.select()中。

- soup.select("a[href='http://example.com/elsie']"):查找 href="http://example.com/elsie" 的<a>节点。
- soup.select("a[href$='sie']"):查找 href 以 sie 结尾的<a>节点。
- soup.select("a[href^='http://example.com']"):查找 href 以"http://example.com"开始的<a>节点。
- soup.select("a[href*='example']"):查找 href 中包含 example 字符串的<a>节点。

2.5.3 使用 soup.select()查找子孙节点

微课 28
Select 查找子孙节点

在 soup.select()中的 CSS 有多个节点时,节点元素之间用空格分开,就是查找子孙节点,例如 soup.select("div p")是查找所有<div>节点下面的所有子孙<p>节点。

例 2-5-3: 查找子孙节点

```
from bs4 import BeautifulSoup
doc="<div><p>A</p><span><p>B</p></span></div><div><p>C</p></div>"
soup=BeautifulSoup(doc,"lxml")
tags=soup.select("div p")
for tag in tags:
    print(tag)
```

程序运行结果:

```
<p>A</p>
<p>B</p>
<p>C</p>
```

tags=soup.select("div p")是查找<div>节点下面所有子孙节点<p>,因此包含<span>节点下面的<p>B</p>。

2.5.4　使用 soup.select()查找直接子节点

在 soup.select()中节点元素之间用 > 分开表示查找所有子节点（注意：>的前后至少包含一个空格），例如 soup.select("div > p")是查找所有<div>节点下面的所有直接子节点<p>，不包含孙节点。

例 2-5-4：查找所有子节点

```
from bs4 import BeautifulSoup
doc="<div><p>A</p><span><p>B</p></span></div><div><p>C</p></div>"
soup=BeautifulSoup(doc,"lxml")
tags=soup.select("div > p")
for tag in tags:
    print(tag)
```

程序运行结果：

```
<p>A</p>
<p>C</p>
```

其中 tags=soup.select("div > p")是查找<div>下面的子节点<p>，因此不包含<span>节点下面的<p>B</p>。

2.5.5　使用 soup.select()查找兄弟节点

在 soup.select()中用 ~ 连接两个节点表示查找前一个节点后面的所有同级别的兄弟节点（注意~前后至少有一个空格），例如 soup.select("div ~ p")查找<div>后面的所有同级别的<p>兄弟节点。

在 soup.select()中用 + 连接两个节点表示查找前一个节点后面的第一个同级别的兄弟节点（注意+前后至少有一个空格）。

例 2-5-5：查找兄弟节点

```
from bs4 import BeautifulSoup
doc="<body>demo<div>A</div><b>X</b><p>B</p><span><p>C</p></span><p>D</p></div></body>"
soup=BeautifulSoup(doc,"lxml")
print(soup.prettify())
tags=soup.select("div ~ p")
for tag in tags:
    print(tag)
print()
tags=soup.select("div + p")
for tag in tags:
    print(tag)
```

程序运行结果：

```
<p>B</p>
<p>D</p>
```

其中 tags=soup.select("div ~ p")找到\<div\>后面同级别的所有\<p\>节点，不包含\<span\>中的\<p\>C\</p\>，因为该节点与\<div\>级别不同。而 tags=soup.select("div + p")需要找\<div\>下一个兄弟节点\<p\>，但\<div\>的下一个兄弟节点是\<b\>X\</b\>，不是\<p\>节点，因此没有找到。注意结果不是\<p\>B\</p\>。

2.6 爬取图书网站数据

在 1.5 节中创建了一个图书网站，现在可以编写一个爬虫程序使用 BeautifulSoup 获取首页数据。

2.6.1 分析网站结构

启动图书网站服务器，在浏览器中输入图书网站网址 http://127.0.0.1:5000，右击页面，在快捷菜单中单击"检查"命令就可以看到网页的结构代码，如图 2-6-1 所示。

微课 29
爬取天气预报数据 1

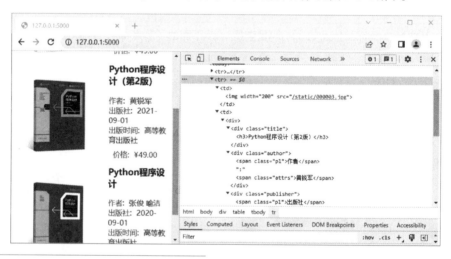

图 2-6-1
网页结构代码

每本图书的记录都包含在一个\<tr\>元素中，\<tr\>中的第一个\<td\>元素包含图书的图片，第二个\<td\>元素包含图书的数据。

2.6.2 获取图书数据

微课 30
爬取天气预报数据 2

1. 获取图书图片

先找到\<table\>中的所有\<tr\>元素，再循环查找每个\<tr\>元素，就可以找到每本图书的图片与文本数据：

```
url="http://127.0.0.1:5000"
resp=urllib.request.urlopen(url)
html=resp.read().decode()
soup=BeautifulSoup(html,"lxml")
trs=soup.select("table tr")
```

2.6 爬取图书网站数据

```
for tr in trs:
    #查找每本图书数据
```

<tr>中的第一个<td>包含图片元素<img>，获取图像 src：

```
src=tr.select_one("td:first-child img")["src"]
src=urllib.request.urljoin(url,src)
```

其中 urljoin()函数把 url 与 src 组合成完整的地址。获取图片地址 src 后就可以设计一个下载图片的函数，把图片下载到 download 文件夹中：

```
def download(src):
    try:
        #获取文件名称
        p=src.rfind("/")
        fn=src[p+1:]
        #读取文件二进制数据
        resp=urllib.request.urlopen(src)
        data=resp.read()
        #保存文件
        f=open("download\\"+fn,"wb")
        f.write(data)
        f.close()
        print("Downloaded",fn)
    except Exception as err:
        print(err)
```

2．获取图书数据

在<tr>元素的最后一个<td>中包含了图书的文本数据，分析网页的结构，很容易获取到图书的 Title（名称）、Author（作者）、Publisher（出版社）、PubDate（出版日期）、Price（价格）等：

```
td=tr.select_one("td:last-child")
Title=td.select_one("div[class='title'] h3").text
Author=td.select_one("div[class='author'] span:last-child").text
Publisher=td.select_one("div[class='publisher'] span:last-child").text
PubDate=td.select_one("div[class='date'] span:last-child").text
Price=td.select_one("div[class='price'] span:last-child").text
```

2.6.3 编写爬虫程序

编写爬虫程序如下：

```
from bs4 import BeautifulSoup
import urllib.request
import os
```

微课 31
爬取天气预报数据 3

```python
def download(src):
    try:
        #获取文件名称
        p=src.rfind("/")
        fn=src[p+1:]
        #读取文件二进制数据
        resp=urllib.request.urlopen(src)
        data=resp.read()
        #保存文件
        f=open("download\\"+fn,"wb")
        f.write(data)
        f.close()
        print("Downloaded",fn)
    except Exception as err:
        print(err)


try:
    #如果 download 不存在就创建
    if not os.path.exists("download"):
        os.mkdir("download")
    url="http://127.0.0.1:5000"
    resp=urllib.request.urlopen(url)
    html=resp.read().decode()
    soup=BeautifulSoup(html,"lxml")
    #获取所有<tr>
    trs=soup.select("table tr")
    for tr in trs:
        #查找每本书籍数据
        src=tr.select_one("td:first-child img")["src"]
        src=urllib.request.urljoin(url,src)
        #下载图像
        download(src)
        td=tr.select_one("td:last-child")
        Title=td.select_one("div[class='title'] h3").text
        Author=td.select_one("div[class='author'] span:last-child").text
        Publisher=td.select_one("div[class='publisher'] span:last-child").text
        PubDate=td.select_one("div[class='date'] span:last-child").text
        Price=td.select_one("div[class='price'] span:last-child").text
        print(Title)
        print(Author)
```

```
            print(Publisher)
            print(PubDate)
            print(Price)
            print()
    except Exception as err:
        print(err)
```

成功爬取首页数据，程序运行结果：

```
Downloaded 000001.jpg
Python 语言程序设计基础（第 2 版）
嵩天、礼欣、黄天羽
2017-02-01
高等教育出版社
¥29.20

Downloaded 000002.jpg
Python 语言程序设计
李学刚

高等教育出版社
¥49.00

Downloaded 000003.jpg
Python 程序设计（第 2 版）
黄锐军
2021-09-01
高等教育出版社
¥49.00
```

成功存储首页的 3 张图片，如图 2-6-2 所示。

图 2-6-2 首页图片

2.7 实践项目——爬取天气预报数据

2.7.1 项目简介

在中国天气网中输入一个城市的名称，例如输入深圳，那么会转到显示深圳的天气预报的网页，每个城市或者地区都有一个代码，如图 2-7-1、图 2-7-2 所示。

图 2-7-1
中国天气网站

图 2-7-2
深圳的天气预报

在图 2-7-2 中可以看到深圳 7 天的天气预报，本任务是爬取 7 天的天气预报数据。

2.7.2 HTML 代码分析

用浏览器浏览网站，在"7 天"选项卡当天天气预报处右击，在快捷菜单中选择"检查"命令，可以打开该位置对应的 HTML 代码，如图 2-7-3 所示。

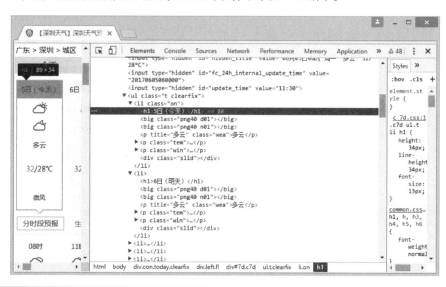

图 2-7-3
HTML 代码

2.7 实践项目——爬取天气预报数据

选择<ul class="t clearfix">元素,右击,在快捷菜单中选择 Edit as HTML 命令,进入编辑状态,复制整个 HTML,结果如下:

```
<ul class="t clearfix">
<li class="on">
<h1>5 日(今天)</h1>
<big class="png40 d01"></big>
<big class="png40 n01"></big>
<p title="多云" class="wea">多云</p>
<p class="tem">
<span>32</span>/<i>28℃</i>
</p>
<p class="win">
<em>
<span title="无持续风向" class=""></span>
<span title="无持续风向" class=""></span>
</em>
<i>微风</i>
</p>
<div class="slid"></div>
</li>
<li>
<h1>6 日(明天)</h1>
<big class="png40 d01"></big>
<big class="png40 n01"></big>
<p title="多云" class="wea">多云</p>
<p class="tem">
<span>32</span>/<i>27℃</i>
</p>
<p class="win">
<em>
<span title="无持续风向" class=""></span>
<span title="无持续风向" class=""></span>
</em>
<i>微风</i>
</p>
<div class="slid"></div>
</li>
<li>
<h1>7 日(后天)</h1>
<big class="png40 d01"></big>
```

```html
<big class="png40 n01"></big>
<p title="多云" class="wea">多云</p>
<p class="tem">
<span>32</span>/<i>27℃</i>
</p>
<p class="win">
<em>
<span title="无持续风向" class=""></span>
<span title="无持续风向" class=""></span>
</em>
<i>微风</i>
</p>
<div class="slid"></div>
</li>
<li>
<h1>8日（周四）</h1>
<big class="png40 d01"></big>
<big class="png40 n01"></big>
<p title="多云" class="wea">多云</p>
<p class="tem">
<span>32</span>/<i>27℃</i>
</p>
<p class="win">
<em>
<span title="无持续风向" class=""></span>
<span title="无持续风向" class=""></span>
</em>
<i>微风</i>
</p>
<div class="slid"></div>
</li>
<li>
<h1>9日（周五）</h1>
<big class="png40 d01"></big>
<big class="png40 n01"></big>
<p title="多云" class="wea">多云</p>
<p class="tem">
<span>33</span>/<i>27℃</i>
</p>
<p class="win">
<em>
```

```
<span title="无持续风向" class=""></span>
<span title="无持续风向" class=""></span>
</em>
<i>微风</i>
</p>
<div class="slid"></div>
</li>
<li>
<h1>10 日（周六）</h1>
<big class="png40 d01"></big>
<big class="png40 n01"></big>
<p title="多云" class="wea">多云</p>
<p class="tem">
<span>33</span>/<i>27℃</i>
</p>
<p class="win">
<em>
<span title="无持续风向" class=""></span>
<span title="无持续风向" class=""></span>
</em>
<i>微风</i>
</p>
<div class="slid"></div>
</li>
<li>
<h1>11 日（周日）</h1>
<big class="png40 d01"></big>
<big class="png40 n07"></big>
<p title="多云转小雨" class="wea">多云转小雨</p>
<p class="tem">
<span>33</span>/<i>26℃</i>
</p>
<p class="win">
<em>
<span title="无持续风向" class=""></span>
<span title="无持续风向" class=""></span>
</em>
<i>微风</i>
</p>
<div class="slid"></div>
</li>
</ul>
```

分析这段代码不难发现 7 天的天气预报实际上在一个<ul class="t clearfix">元素中，每天的天气预报都是一个<li>元素，而且结构一样。因此可以通过 BeautifulSoup 的元素查找方法得到各元素的值。

2.7.3 爬取天气预报数据

通过分析 HTML 代码，可以编写程序爬取深圳 7 天的天气预报数据。程序如下：

```python
from bs4 import BeautifulSoup
from bs4 import UnicodeDammit
import urllib.request

url="http://www.weather.com.cn/weather/101280601.shtml"
try:
    headers={"User-Agent":"Mozilla/5.0 (Windows; U; Windows NT 6.0 x64; en-US; rv:1.9pre) Gecko/2008072421 Minefield/3.0.2pre"}
    req=urllib.request.Request(url,headers=headers)
    data=urllib.request.urlopen(req)
    data=data.read()
    dammit=UnicodeDammit(data,["utf-8","gbk"])
    data=dammit.unicode_markup
    soup=BeautifulSoup(data,"lxml")
    lis=soup.select("ul[class='t clearfix'] li")
    for li in lis:
        try:
            date=li.select('h1')[0].text
            weather=li.select('p[class="wea"]')[0].text
            temp=li.select('p[class="tem"] span')[0].text+"/"+li.select('p[class="tem"] i')[0].text
            print(date,weather,temp)
        except Exception as err:
            print(err)
except Exception as err:
    print(err)
```

程序爬取结果：

```
5日（今天）   多云 32℃/28℃
6日（明天）   多云 32℃/27℃
7日（后天）   多云 32℃/27℃
8日（周四）   多云 32℃/27℃
9日（周五）   多云 33℃/27℃
10日（周六）  多云 33℃/27℃
11日（周日）  多云转小雨 33℃/26℃
```

可见爬取数据与在网站上显示的数据一样。

2.7.4 爬取与存储天气预报数据

获取北京、上海、广州、深圳等城市的代码，爬取这些城市的天气预报数据，并存储到 sqllite 数据库 weathers.db 中，存储的数据表 weathers 是：

> create table weathers (wCity varchar(16),wDate varchar(16),wWeather varchar(64),wTemp varchar(32),constraint pk_weather primary key (wCity,wDate))"

编写程序依次爬取各个城市的天气预报数据并存储在数据库中。程序如下：

```python
from bs4 import BeautifulSoup
from bs4 import UnicodeDammit
import urllib.request
import sqlite3

class WeatherDB:
    def openDB(self):
        self.con=sqlite3.connect("weathers.db")
        self.cursor=self.con.cursor()
        try:
            self.cursor.execute("create table weathers (wCity varchar(16),wDate varchar(16), wWeather varchar(64),wTemp varchar(32),constraint pk_weather primary key (wCity,wDate))")
        except:
            self.cursor.execute("delete from weathers")

    def closeDB(self):
        self.con.commit()
        self.con.close()

    def insert(self,city,date,weather,temp):
        try:
            self.cursor.execute("insert into weathers (wCity,wDate,wWeather,wTemp) values (?,?,?,?)" ,(city,date,weather,temp))
        except Exception as err:
            print(err)

    def show(self):
        self.cursor.execute("select * from weathers")
        rows=self.cursor.fetchall()
        print("%-16s%-16s%-32s%-16s" % ("city","date","weather","temp"))
```

```python
            for row in rows:
                print("%-16s%-16s%-32s%-16s" % (row[0],row[1],row[2],row[3]))

    class WeatherForecast:
        def __init__(self):
            self.headers = {
                "User-Agent": "Mozilla/5.0 (Windows; U; Windows NT 6.0 x64; en-US; rv:1.9pre) Gecko/2008072421 Minefield/3.0.2pre"}
            self.cityCode={"北京":"101010100","上海":"101020100","广州":"101280101","深圳":"101280601"}

        def forecastCity(self,city):
            if city not in self.cityCode.keys():
                print(city+" code cannot be found")
                return

            url="http://www.weather.com.cn/weather/"+self.cityCode[city]+".shtml"
            try:
                req=urllib.request.Request(url,headers=self.headers)
                data=urllib.request.urlopen(req)
                data=data.read()
                dammit=UnicodeDammit(data,["utf-8","gbk"])
                data=dammit.unicode_markup
                soup=BeautifulSoup(data,"lxml")
                lis=soup.select("ul[class='t clearfix'] li")
                for li in lis:
                    try:
                        date=li.select('h1')[0].text
                        weather=li.select('p[class="wea"]')[0].text
                        temp=li.select('p[class="tem"] span')[0].text+"/"+li.select('p[class="tem"] i')[0].text
                        print(city,date,weather,temp)
                        self.db.insert(city,date,weather,temp)
                    except Exception as err:
                        print(err)
            except Exception as err:
                print(err)

        def process(self,cities):
            self.db=WeatherDB()
            self.db.openDB()
```

```
        for city in cities:
            self.forecastCity(city)

        #self.db.show()
        self.db.closeDB()

ws=WeatherForecast()
ws.process(["北京","上海","广州","深圳"])
print("completed")
```

程序运行结果：

```
北京 7 日（今天） 晴间多云，北部山区有阵雨或雷阵雨转晴转多云 31℃/17℃
北京 8 日（明天） 多云转晴，北部地区有分散阵雨或雷阵雨转晴 34℃/20℃
北京 9 日（后天） 晴转多云 36℃/22℃
北京 10 日（周六） 阴转阵雨 30℃/19℃
北京 11 日（周日） 阵雨 27℃/18℃
北京 12 日（周一） 阴转晴 28℃/20℃
北京 13 日（周二） 晴 32℃/21℃
上海 7 日（今天） 多云 30℃/21℃
上海 8 日（明天） 多云转阴 32℃/23℃
上海 9 日（后天） 阵雨 32℃/24℃
上海 10 日（周六） 中雨 27℃/22℃
上海 11 日（周日） 小雨转多云 29℃/22℃
上海 12 日（周一） 多云 30℃/22℃
上海 13 日（周二） 多云转阴 30℃/21℃
广州 7 日（今天） 多云 35℃/27℃
广州 8 日（明天） 多云 35℃/28℃
广州 9 日（后天） 多云 35℃/28℃
广州 10 日（周六） 多云 35℃/28℃
广州 11 日（周日） 多云 35℃/28℃
广州 12 日（周一） 雷阵雨 35℃/27℃
广州 13 日（周二） 雷阵雨转大雨 33℃/24℃
深圳 7 日（今天） 阵雨转多云 34℃/28℃
深圳 8 日（明天） 晴 34℃/28℃
深圳 9 日（后天） 晴 34℃/28℃
深圳 10 日（周六） 晴转阵雨 34℃/28℃
深圳 11 日（周日） 阵雨 33℃/27℃
深圳 12 日（周一） 阵雨 32℃/27℃
深圳 13 日（周二） 阵雨转中雨 32℃/25℃
```

练习二

① 简单说明 BeautifulSoup 解析数据的特点。

② 用 BeautifulSoup 装载以下 HTML 文档，打印规范格式，与原来 HTML 文档比较区别，说明 BeautifulSoup 是如何修改的？

```
<body>
<div>Hi<br>
<span>Hello</SPAN>
```

③ 重新编写项目 1 中爬取学生信息的程序，使用 BeautifulSoup 分解出<tr>...</tr>中的<td>...</td>数据，补全以下 searchWeb 函数。

```
def searchWeb(html):
    rows=[]
    soup=BeautifulSoup(html,"lxml")
    trs=soup.find_all("tr")
    for i in _____:
        row=[]
        tds=trs[i].find_all("td")
        for td in tds:
            _____
        rows.append(row)
    return rows
```

④ 以下是一段 HTML 代码：

```
<body>
<bookstore>
<book id="b1">
  <title lang="English">Harry Potter</title>
  <price>23.99</price>
</book>
<book id="b2">
  <title lang="Chinese">学习 XML</title>
  <price>39.95</price>
</book>
<book id="b3">
  <title lang="English">Learning Python</title>
  <price>30.20</price>
</book>
```

```
        </bookstore>
    </body></html>
```

使用 BeautifulSoup 完成以下任务：
- 找出所有英文书的名称与价格。
- 找出所有价格在 30 元以上的书的名称。

⑤ 编写一个爬取 Python 程序最新版本的程序，程序分为两部分。
- 访问 Python 官方网站的下载页面，如图 2-8-1 所示。

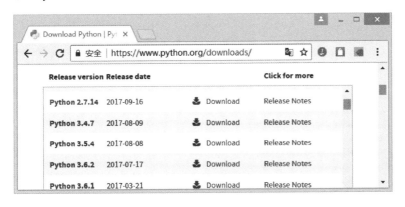

图 2-8-1
Python 官网的下载页面

分析 HTML 代码结构，可以编写以下程序爬取所有发行的 Python 版本与下载地址。

```
from bs4 import    BeautifulSoup
import urllib.request
def searchPython(url):
    resp=urllib.request.urlopen(url)
    data=resp.read()
    html=data.decode()
    soup=BeautifulSoup(html,"lxml")
    ol=soup.find(name="ol",attrs={"class":"list-row-container menu"})
    lis=ol.find_all("li")
    for li in lis:
        a=li.find(name="span",attrs={"class":"release-number"}).find("a")
        python=a.text
        url=a["href"]
        print("%-20s %s" %(python,url))
try:
    searchPython("https://www.python.org/downloads/")
except Exception as e:
    print(e)
```

程序运行结果：

```
Python 2.5.4            /download/releases/2.5.4/
```

	Python 2.4.6	/download/releases/2.4.6/
	Python 2.5.3	/download/releases/2.5.3/
	Python 2.6.1	/download/releases/2.6.1/
	Python 3.0.0	/download/releases/3.0/
	……	

- 在该程序的基础上继续编写，找出最新的版本，并自动进入最新版本下载页面，找出这个版本的 Windows 64 位的 ZIP 压缩包的下载地址，自动下载该 Python 压缩包。

项目 3　爬取旅游网站数据

本项目介绍如何使用深度优先与广度优先的策略来爬取网站中多个网页的数据的方法，同时介绍多线程与分布式爬取数据的方法，最后介绍爬取旅游网站图片文件的综合案例。

3.1 网站树的爬取路径

一个网站往往由很多相互关联的网页组成,每个网页上都可能包含所需数据,那么如何获取这些数据呢?显然必须穿梭于各个网页之间,那么按什么样的规则穿梭呢?常用的策略有深度优先与广度优先方法。为了讲解这两种方法的工作过程,本节特意设计一个简单的网站。

3.1.1 Web 服务器网站

设计好 books.htm、program.htm、database.htm、netwwork.htm、mysql.htm、java.htm、python.htm 等网页文件以 UTF-8 编码格式存储在文件夹中,各个文件的内容如下:

(1) books.htm

```
<h3>计算机</h3>
<ul>
<li><a href="database.htm">数据库</a></li>
<li><a href="program.htm">程序设计</a></li>
<li><a href="network.htm">计算机网络</a></li>
</ul>
```

(2) database.htm

```
<h3>数据库</h3>
<ul>
<li><a href="mysql.htm">MySQL 数据库</a></li>
</ul>
```

(3) program.htm

```
<h3>程序设计</h3>
<ul>
<li><a href="python.htm">Python 程序设计</a></li>
<li><a href="java.htm">Java 程序设计</a></li>
</ul>
```

(4) network.htm

```
<h3>计算机网络</h3>
```

(5) mysql.htm

```
<h3>MySQL 数据库</h3>
```

(6) python.htm

```
<h3>Python 程序设计</h3>
```

(7) java.htm

```
<h3>Java 程序设计</h3>
```

然后用 Flask 设计一个 server.py 的 Web 程序来呈现这些网页：

```python
import flask
import os
app=flask.Flask(__name__)

def getFile(fileName):
    data=b""
    if os.path.exists(fileName):
        fobj=open(fileName,"rb")
        data=fobj.read()
        fobj.close()
    return data

@app.route("/")
def index():
    return getFile("books.htm")

@app.route("/<section>")
def process(section):
    data=""
    if section!="":
        data=getFile(section)
    return data

if __name__=="__main__":
    app.run()
```

该程序运行后 Web 默认网址是 http://127.0.0.1:5000，访问这个网址后自动执行 index() 函数，返回 books.htm 网页，如图 3-1-1 所示。

图 3-1-1 books.htm 网页

单击每个超链接后会转到数据库、程序设计、计算机网络各个网页，这些网页的结构实际上是一棵树，如图 3-1-2 所示。

图 3-1-2 网页结构

3.1.2 使用递归程序爬取数据

微课 33
递归程序爬取数据

设计一个客户端程序 client.py 爬取这个网站各个网页的<h3>的标题值，设计的思路如下：

① 从 books.htm 出发。
② 访问一个网页，获取<h3>标题。
③ 获取这个网页中所有<a>超链接的 href 值形成 links 列表。
④ 循环遍历 links 列表，对于每个链接 link 都指向一个网页，递归回到②。
⑤ 继续 links 列表中的下一个 link，直到遍历完列表中所有 link。

因此该程序是一个递归调用的过程，程序如下：

```python
from bs4 import BeautifulSoup
import urllib.request

def spider(url):
    try:
        data=urllib.request.urlopen(url)
        data=data.read()
        data=data.decode()
        soup=BeautifulSoup(data,"lxml")
        print(soup.find("h3").text)
        links=soup.select("a")
        for link in links:
            href=link["href"]
            url=start_url+"/"+href
            #print(url)
            spider(url)
    except Exception as err:
        print(err)

start_url="http://127.0.0.1:5000"
spider(start_url)
print("The End")
```

程序运行结果：

```
计算机
数据库
MySQL 数据库
程序设计
Python 程序设计
Java 程序设计
计算机网络
The End
```

如果读者熟悉数据结构，可以发现程序在使用深度优先方式遍历这棵树，实际上这种递归程序都是采用深度优先的方式遍历树的。

3.1.3 使用深度优先策略爬取数据

如果不使用递归方式实现深度优先的策略爬取网站数据，也可以设计一个栈来完成。实现一个栈十分简单，Python 中的列表就是一个栈。编写代码如下：

微课 34
深度优先爬取数据

```python
class Stack:
    def __init__(self):
        self.st=[]
    def pop(self):
        return self.st.pop()
    def push(self,obj):
        self.st.append(obj)
    def empty(self):
        return len(self.st)==0
```

push()是进栈函数，pop()是出栈函数，empty()判断栈是否为空。使用 Stack 类后可以设计深度优先的策略爬取数据的客户端程序，思路如下：

① 第一个 URL 进栈。
② 如果栈为空则程序结束，如不为空则出栈一个 URL，并爬取该网页的<h3>标题值。
③ 获取 URL 站点的所有超链接<a>的 href 值，组成 links 链接列表，把这些链接全部进栈。
④ 回到②。

编写 client.py 程序如下：

```python
from bs4 import BeautifulSoup
import urllib.request

class Stack:
    def __init__(self):
        self.st=[]
    def pop(self):
        return self.st.pop()
    def push(self,obj):
        self.st.append(obj)
    def empty(self):
        return len(self.st)==0

def spider(url):
    stack=Stack()
    stack.push(url)
    while not stack.empty():
```

```
            url=stack.pop()
            try:
                data=urllib.request.urlopen(url)
                data=data.read()
                data=data.decode()
                soup=BeautifulSoup(data,"lxml")
                print(soup.find("h3").text)
                links=soup.select("a")
                for i in range(len(links)-1,-1,-1):
                    href=links[i]["href"]
                    url=start_url+"/"+href
                    stack.push(url)
            except Exception as err:
                print(err)

start_url="http://127.0.0.1:5000"
spider(start_url)
print("The End")
```

程序运行结果：

```
计算机
数据库
MySQL 数据库
程序设计
Python 程序设计
Java 程序设计
计算机网络
The End
```

微课 35
广度优先爬取数据

3.1.4 广度优先策略爬取数据

遍历网站树还可使用广度优先的策略，可以使用队列实现，在 Python 中实现队列十分简单，Python 中的列表就是一个队列，编写代码如下：

```
class Queue:
    def __init__(self):
        self.st=[]
    def fetch(self):
        return self.st.pop(0)
    def enter(self,obj):
        self.st.append(obj)
    def empty(self):
```

```
            return len(self.st)==0
```

enter()是入列函数、fetch()是出列函数、empty()判断列是否为空。使用 Queue 类后可以设计广度优先的策略爬取数据的客户端程序,思路如下:

① 第一个 URL 入列。
② 如果列为空则程序结束,如不为空,则出列一个 URL,爬取该网页的<h3>标题值。
③ 获取该 URL 网页的所有超链接<a>的 href 值,组成 links 链接列表,把这些链接全部入列。
④ 回到②。

client.py 程序如下:

```
from bs4 import BeautifulSoup
import urllib.request

class Queue:
    def __init__(self):
        self.st=[]
    def fetch(self):
        return self.st.pop(0)
    def enter(self,obj):
        self.st.append(obj)
    def empty(self):
        return len(self.st)==0


def spider(url):
    queue=Queue()
    queue.enter(url)
    while not queue.empty():
        url=queue.fetch()
        try:
            data=urllib.request.urlopen(url)
            data=data.read()
            data=data.decode()
            soup=BeautifulSoup(data,"lxml")
            print(soup.find("h3").text)
            links=soup.select("a")
            for link in links:
                href=link["href"]
                url=start_url+"/"+href
                queue.enter(url)
        except Exception as err:
            print(err)
```

```
                start_url="http://127.0.0.1:5000"
                spider(start_url)
                print("The End")
```

程序运行结果：

```
                计算机
                数据库
                程序设计
                计算机网络
                MySQL 数据库
                Python 程序设计
                Java 程序设计
                The End
```

3.2 网站图的爬取路径

PPT 复杂的 Web 网站

微课 36
复杂的 Web 网站

深度优先与广度优先策略都是遍历树的方法，但是网站的各网页之间的关系未必是树结构，可能组成一个复杂的结构，可能存在回路。如果在网站中每个网页都加一条<a href="books.htm">Home</a>的语句，让每个网页都能回到主界面，那么网站的关系就是一个有回路的图。

3.2.1 复杂的 Web 网站

（1）books.htm

```
                <h3>计算机</h3>
                <ul>
                <li><a href="database.htm">数据库</a></li>
                <li><a href="program.htm">程序设计</a></li>
                <li><a href="network.htm">计算机网络</a></li>
                </ul>
```

（2）database.htm

```
                <h3>数据库</h3>
                <ul>
                <li><a href="mysql.htm">MySQL 数据库</a></li>
                </ul>
                <a href="books.htm">Home</a>
```

（3）program.htm

```
                <h3>程序设计</h3>
                <ul>
                <li><a href="python.htm">Python 程序设计</a></li>
                <li><a href="java.htm">Java 程序设计</a></li>
```

```
</ul>
<a href="books.htm">Home</a>
```

（4）network.htm

```
<h3>计算机网络</h3>
<a href="books.htm">Home</a>
```

（5）mysql.htm

```
<h3>MySQL 数据库</h3>
<a href="books.htm">Home</a>
```

（6）python.htm

```
<h3>Python 程序设计</h3>
<a href="books.htm">Home</a>
```

（7）java.htm

```
<h3>Java 程序设计</h3>
<a href="books.htm">Home</a>
```

这时，深度优先与广度优先策略方法要做改进，使用 Python 中的列表 urls 来存储访问过的网站。如果一个网址 URL 没在该列表中就可以访问，并把网站的 URL 加到 urls 中保存。如果 URL 在该列表中就不访问。这样可以避免形成回路，导致无限循环。

3.2.2 改进客户端深度优先策略程序

图 G 的初始状态的顶点均未曾访问。在 G 中任选一顶点 v 为初始出发点（源点），则深度优先遍历的步骤如下：首先访问出发点 v，并将其标记为已访问；然后依次从 v 出发搜索 v 的每个邻接点 w。若 w 未曾访问，则以 w 为新的出发点继续进行深度优先遍历，直至图中所有和源点 v 有相通路径的顶点（亦称为从源点可达的顶点）均已被访问为止。

图的深度优先遍历类似于树的前序遍历。采用的搜索方法的特点是尽可能先对纵深方向进行搜索。这种搜索方法称为深度优先搜索（Depth-First Search）。相应地，用此方法遍历图就很自然地被称为图的深度优先遍历，基本实现思路：

① 访问顶点 v。
② 从 v 的未被访问的邻接点中选取一个顶点 w，从 w 出发进行深度优先遍历。
③ 重复上述两步，直至图中所有和 v 有相通路径的顶点都被访问为止。
（1）使用递归实现，编写程序如下：

微课 37
改进深度优先客户端
程序

```
from bs4 import BeautifulSoup
import urllib.request

def spider(url):
    global urls
    if url not in urls:
        urls.append(url)
```

```
                    try:
                        data=urllib.request.urlopen(url)
                        data=data.read()
                        data=data.decode()
                        soup=BeautifulSoup(data,"lxml")
                        print(soup.find("h3").text)
                        links=soup.select("a")
                        for link in links:
                            href=link["href"]
                            url=start_url+"/"+href
                            spider(url)
                    except Exception as err:
                        print(err)


start_url="http://127.0.0.1:5000"
urls=[]
spider(start_url)
print("The End")
```

(2)使用栈实现，编写程序如下：

```
from bs4 import BeautifulSoup
import urllib.request

class Stack:
    def __init__(self):
        self.st=[]
    def pop(self):
        return self.st.pop()
    def push(self,obj):
        self.st.append(obj)
    def empty(self):
        return len(self.st)==0

def spider(url):
    global urls
    stack=Stack()
    stack.push(url)
    while not stack.empty():
        url=stack.pop()
        if url not in urls:
            urls.append(url)
```

```
                try:
                    data=urllib.request.urlopen(url)
                    data=data.read()
                    data=data.decode()
                    soup=BeautifulSoup(data,"lxml")
                    print(soup.find("h3").text)
                    links=soup.select("a")
                    for i in range(len(links)-1,-1,-1):
                        href=links[i]["href"]
                        url=start_url+"/"+href
                        stack.push(url)
                except Exception as err:
                    print(err)

start_url="http://127.0.0.1:5000"
urls=[]
spider(start_url)
print("The End")
```

两种实现方法程序运行结果相同：

```
计算机
数据库
MySQL 数据库
计算机
程序设计
Python 程序设计
Java 程序设计
计算机网络
The End
```

3.2.3 改进客户端广度优先策略程序

图的广度优先遍历算法是一个分层搜索的过程，和树的层序遍历算法类同，也需要一个队列以存储遍历过的顶点顺序，以便按出列的顺序再去访问这些顶点的邻接顶点。基本实现思想：

① 顶点 v 入列。
② 当队列非空时则继续执行，否则算法结束。
③ 出列取得队头顶点 v，访问顶点 v 并标记顶点 v 已被访问。
④ 查找顶点 v 的第一个邻接顶点 col。
⑤ 若 v 的邻接顶点 col 未被访问过，则 col 入列。
⑥ 继续查找顶点 v 的另一个新的邻接顶点 col，回到⑤。顶点 v 的所有未被访问过的邻接点处理完。回到②。

微课 38
改进广度优先客户端
程序

广度优先遍历是以顶点 v 为起始点，由近至远，依次访问和 v 有路径相通而且按路径长度升序访问。为了使"先被访问顶点的邻接点"先于"后被访问顶点的邻接点"被访问，须设置队列存储访问的顶点。

编写代码如下：

```python
from bs4 import BeautifulSoup
import urllib.request

class Queue:
    def __init__(self):
        self.st=[]
    def fetch(self):
        return self.st.pop(0)
    def enter(self,obj):
        self.st.append(obj)
    def empty(self):
        return len(self.st)==0

def spider(url):
    global urls
    queue=Queue()
    queue.enter(url)
    while not queue.empty():
        url=queue.fetch()
        if url not in urls:
            try:
                urls.append(url)
                data=urllib.request.urlopen(url)
                data=data.read()
                data=data.decode()
                soup=BeautifulSoup(data,"lxml")
                print(soup.find("h3").text)
                links=soup.select("a")
                for link in links:
                    href=link["href"]
                    url=start_url+"/"+href
                    queue.enter(url)
            except Exception as err:
                print(err)

start_url="http://127.0.0.1:5000"
```

```
            urls=[]
            spider(start_url)
            print("The End")
```

程序运行结果：

```
            计算机
            数据库
            程序设计
            计算机网络
            MySQL 数据库
            计算机
            Python 程序设计
            Java 程序设计
            The End
```

3.3 Python 实现多线程

线程类似于同时执行多个不同程序，多线程运行有如下优点：
- 使用线程可以把程序中运行时间长的任务放到后台处理。
- 程序的运行速度可能加快。
- 在一些需要一段时间等待的任务如用户输入、文件读写和网络收发数据等，多线程使用效果理论上比单线程好。
- 每个线程都有一组 CPU 寄存器，称为线程的上下文，该上下文反映了线程上次运行该线程的 CPU 寄存器的状态。
- 在其他线程正在运行时，线程可以暂时搁置（也称睡眠），又称线程的退让。

PPT　Python 的前后台线程

微课 39
Python 的前后台线程

3.3.1 Python 的前后台线程

在 Python 中要启动一个线程，可以使用 threading 包中的 Thread 建立一个对象。Thread 类的使用方法：

```
            t=Thread(target,args=None)
```

target 是要执行的函数，args 是一个元组或者为 target 的函数提供参数的列表，然后调用 t.start()就启动了线程。

例 3-3-1：在主线程中启动一个子线程执行 reading 函数

```
            import threading
            import time
            import random

            def reading():
                for i in range(10):
                    print("reading",i)
                    time.sleep(random.randint(1,2))
```

```
r=threading.Thread(target=reading)
r.setDaemon(False)
r.start()
print("The End")
```

程序运行结果：

```
reading 0
The End
reading 1
reading 2
reading 3
reading 4
```

主线程启动子线程 r 后运行结束，但是子线程还没有运行结束，继续显示完 reading 4 后才结束。其中的 r.setDaemon(False)就是设置线程 r 为后台线程，后台线程不因主线程的结束而结束。如果设置 r.setDaemon(True)，那么 r 就是前台线程。

例 3-3-2： 启动一个前台线程

```
import threading
import time
import random

def reading():
    for i in range(5):
        print("reading",i)
        time.sleep(random.randint(1,2))

r=threading.Thread(target=reading)
r.setDaemon(True)
r.start()
print("The End")
```

程序运行结果：

```
reading 0
The End
```

可见在主线程结束后子线程也结束，这就是前台线程。

例 3-3-3： 前台与后台线程

```
import threading
import time
import random
```

```
def reading():
    for i in range(5):
        print("reading",i)
        time.sleep(random.randint(1,2))

def test():
    r=threading.Thread(target=reading)
    r.setDaemon(True)
    r.start()
    print("test end")

t=threading.Thread(target=test)
t.setDaemon(False)
t.start()
print("The End")
```

程序运行结果：

```
The End
reading 0
test end
```

可见主线程启动后台子线程 t 后就结束了，但是 t 还在执行，在 t 中启动前台 r 子线程，之后 t 结束，相应的 r 也结束了。

3.3.2 线程的等待

在多线程的程序中往往需要一个线程（如主线程）要等待其他线程执行完毕才继续执行，这时可以使用 join 函数，使用的方法如下：

线程对象.join()

在一个线程代码中执行这条语句，当前的线程就会停止执行，一直等到指定的线程对象的线程执行完毕后才继续执行，这条语句有启动阻塞等待的作用。

例 3-3-4： 主线程启动一个子线程并等待子线程结束后才继续执行

```
import threading
import time
import random

def reading():
    for i in range(5):
        print("reading",i)
        time.sleep(random.randint(1,2))

t=threading.Thread(target=reading)
```

```
t.setDaemon(False)
t.start()
t.join()
print("The End")
```

程序运行结果：

```
reading 0
reading 1
reading 2
reading 3
reading 4
The End
```

可见主线程启动子线程 t 执行 reading() 函数，t.join() 阻塞了主线程，等到 t 线程执行完毕后 t.join() 执行完毕，继续执行最后输出 The End。

例 3-3-5： 在一个子线程启动另外一个子线程，并等待子线程结束后才继续执行

```
import threading
import time
import random

def reading():
    for i in range(5):
        print("reading",i)
        time.sleep(random.randint(1,2))

def test():
    r=threading.Thread(target=reading)
    r.setDaemon(True)
    r.start()
    r.join()
    print("test end")

t=threading.Thread(target=test)
t.setDaemon(False)
t.start()
t.join()
print("The End")
```

程序运行结果：

```
reading 0
reading 1
reading 2
```

```
reading 3
reading 4
test end
The End
```

由此可见主线程启动 t 线程后 t.join()会等待 t 线程结束，在 test()中再次启动 r 子线程，而且 r.join()而阻塞 t 线程，等待 r 线程执行完毕后才结束 r.join()，然后在 test()中输出 test end，之后 t 线程结束，t.join()执行完毕，主线程中输出 The End 后结束。

3.3.3 多线程与资源

在多个线程的程序中存在一个问题：如果多个子线程要同时访问与改写公共资源，那么应该怎么样协调各个线程。一个解决方法是使用线程锁，Python 使用 threading.RLock 类来创建一个线程锁对象，使用方法如下：

微课 41
多线程与资源 1

```
lock=threading.RLock()
```

lock 对象有两个重要方法 acquire()与 release()，执行 lock.acquire()语句时强迫 lock 获取线程锁。如果已经有另外的线程先调用了 acquire()方法获取了线程锁而还没有调用 release()释放锁，那么这个 lock.acquire()就阻塞当前的线程，一直等待锁的控制权，直到别的线程释放锁后 lock.acquire()就获取锁并解除阻塞，线程继续执行，执行后线程要调用 lock.release()释放锁，不然别的线程会一直得不到锁的控制权导致线程阻塞。

使用 acquire()和 release()的工作机制把一段修改公共资源的代码放在 acquire()与 release()中间，这样就保证一次最多只有一个线程在修改公共资源，别的线程如果也要修改就必须等待，直到占用线程调用 release()释放锁后其他线程才能获取锁的控制权进行资源的修改。

微课 42
多线程与资源 2

例 3-3-6： 子线程 A 把一个全局列表 words 进行升序的排列，子线程 D 把 words 列表进行降序的排列

```python
import threading
import time

lock=threading._RLock()
words=["a","b","d","b","p","m","e","f","b"]

def increase():
    global words
    for count in range(5):
        lock.acquire()
        print("A acquired")
        for i in range(len(words)):
            for j in range(i+1,len(words)):
                if words[j]<words[i]:
                    t=words[i]
                    words[i]=words[j]
```

```python
                        words[j]=t
                print("A ",words)
                time.sleep(1)
                lock.release()

        def decrease():
            global words
            for count in range(5):
                lock.acquire()
                print("D acquired")
                for i in range(len(words)):
                    for j in range(i+1,len(words)):
                        if words[j]>words[i]:
                            t=words[i]
                            words[i]=words[j]
                            words[j]=t
                print("D ",words)
                time.sleep(1)
                lock.release()

        A=threading.Thread(target=increase)
        A.setDaemon(False)
        A.start()

        D=threading.Thread(target=decrease)
        D.setDaemon(False)
        D.start()
        print("The End")
```

程序运行结果：

```
A acquired
A   ['a', 'b', 'b', 'b', 'd', 'e', 'f', 'm', 'p']
The End
D acquired
D   ['p', 'm', 'f', 'e', 'd', 'b', 'b', 'b', 'a']
D acquired
D   ['p', 'm', 'f', 'e', 'd', 'b', 'b', 'b', 'a']
A acquired
A   ['a', 'b', 'b', 'b', 'd', 'e', 'f', 'm', 'p']
A acquired
A   ['a', 'b', 'b', 'b', 'd', 'e', 'f', 'm', 'p']
```

```
D acquired
D ['p', 'm', 'f', 'e', 'd', 'b', 'b', 'b', 'a']
D acquired
D ['p', 'm', 'f', 'e', 'd', 'b', 'b', 'b', 'a']
D acquired
D ['p', 'm', 'f', 'e', 'd', 'b', 'b', 'b', 'a']
A acquired
A ['a', 'b', 'b', 'b', 'd', 'e', 'f', 'm', 'p']
A acquired
A ['a', 'b', 'b', 'b', 'd', 'e', 'f', 'm', 'p']
```

可见无论是子线程执行升序还是降序，都是在获得锁的控制权的前提下进行的，因此排序过程中另外一个线程必然处于等待状态，不会干扰本次的排序，因此每次输出的结果不受干扰。

如果不使用锁，那么在执行升序的子线程运行时执行降序的子线程也在工作，输出结果既不是升序也不是降序，程序不使用锁的运行结果：

```
The End
A ['p', 'm', 'f', 'e', 'd', 'b', 'b', 'b', 'a']
D ['b', 'p', 'm', 'f', 'e', 'd', 'b', 'b', 'a']
D ['a', 'b', 'e', 'p', 'm', 'f', 'd', 'b', 'b']
A ['p', 'm', 'f', 'e', 'd', 'b', 'b', 'b', 'a']
D ['e', 'p', 'm', 'f', 'd', 'b', 'b', 'b', 'a']
D ['a', 'b', 'p', 'm', 'f', 'e', 'd', 'b', 'b']
A ['p', 'm', 'f', 'e', 'd', 'b', 'b', 'b', 'a']
D ['f', 'p', 'm', 'e', 'd', 'b', 'b', 'b', 'a']
A ['a', 'b', 'b', 'b', 'd', 'e', 'f', 'm', 'p']
A ['a', 'b', 'b', 'b', 'd', 'e', 'f', 'm', 'p']
```

3.4 爬取网站复杂数据

PPT Web 服务器网站

3.4.1 Web 服务器网站

在 Web 网站 mysql.htm、java.htm、python.htm 加上图形并丰富内容：

（1）mysql.htm

```
<h3>MySQL 数据库</h3>
<div>
    MySQL 是一个关系数据库管理系统，由瑞典 MySQL AB 公司开发，目前属于 Oracle 旗下产品。MySQL 是最流行的关系数据库管理系统之一，在 Web 应用方面，MySQL 是最好的 RDBMS (Relational Database Management System，关系数据库管理系统) 应用软件。
</div>
<div>
```

微课 43
Web 服务器网站

```
<img src="mysql.jpg"  />
</div>
<a href="books.htm">Home</a>
```

(2) java.htm

```
<h3>Java 程序设计</h3>
<div>
    Java 是一门面向对象编程语言，不仅吸收了 C++语言的各种优点，还摒弃了 C++
里难以理解的多继承、指针等概念，因此 Java 语言具有功能强大和简单易用两个特征。
Java 语言作为静态面向对象编程语言的代表，极好地实现了面向对象理论，允许程序员
以优雅的思维方式进行复杂的编程。
</div>
<div>
   <img src="java.jpg" />
</div>
<a href="books.htm">Home</a>
```

(3) python.htm

```
<h3>Python 程序设计</h3>
<div>
    Python （英国发音：/'paɪθən/ 美国发音：/'paɪθɑːn/），是一种面向对象的解释型计
算机程序设计语言，由荷兰人 Guido van Rossum 于 1989 年发明，第一个公开发行版发
行于 1991 年。
</div>
<div>
<img src="python.jpg" />
</div>
<a href="books.htm">Home</a>
```

3.4.2 爬取网站的复杂数据

微课 44
爬取网站的复杂数据

接下来爬取 mysql.htm、java.htm、python.htm 网站的简介与图像。简介在网页的第一
个<div>中，图像在<img>中，只有这 3 个网页有这样的特征，于是编写客户端程序如下：

```
from bs4 import BeautifulSoup
import urllib.request

def spider(url):
    global urls
    if url not in urls:
        urls.append(url)
        try:
            data=urllib.request.urlopen(url)
```

```
                    data=data.read()
                    data=data.decode()
                    soup=BeautifulSoup(data,"lxml")
                    print(soup.find("h3").text)
                    divs=soup.select("div")
                    imgs=soup.select("img")
                    if len(divs)>0 and len(imgs)>0:
                        print(divs[0].text)
                        url=start_url+"/"+imgs[0]["src"]
                        urllib.request.urlretrieve(url,"downloaded "+imgs[0]["src"])
                        print("downloaded ",imgs[0]["src"])
                    links=soup.select("a")
                    for link in links:
                        href=link["href"]
                        url=start_url+"/"+href
                        spider(url)
            except Exception as err:
                print(err)

start_url="http://127.0.0.1:5000"
urls=[]
spider(start_url)
print("The End")
```

程序运行结果：

```
计算机
数据库
MySQL 数据库


MySQL 是一个关系数据库管理系统，由瑞典 MySQL AB 公司开发，目前属于 Oracle
旗下产品。MySQL 是最流行的关系数据库管理系统之一，在 Web 应用方面，MySQL
是最好的 RDBMS (Relational Database Management System，关系数据库管理系统) 应用
软件。


downloaded   mysql.jpg
计算机
程序设计
Python 程序设计


Python（英国发音：/ˈpaɪθən/ 美国发音：/ˈpaɪθɑːn/），是一种面向对象的解释型计算
机程序设计语言，由荷兰人 Guido van Rossum 于 1989 年发明，第一个公开发行版发行于
```

1991 年。

downloaded   python.jpg
Java 程序设计

Java 是一门面向对象编程语言，不仅吸收了 C++语言的各种优点，还摒弃了 C++里难以理解的多继承、指针等概念，因此 Java 语言具有功能强大和简单易用两个特征。Java 语言作为静态面向对象编程语言的代表，极好地实现了面向对象理论，允许程序员以优雅的思维方式进行复杂的编程。

downloaded   java.jpg
计算机网络
The End

程序运行完毕可以看到 3 个下载完成的图片文件。

其中程序中的以下部分：

```
if len(divs) > 0 and len(imgs) > 0:
    print(divs[0].text)
    url = start_url + "/" + imgs[0]["src"]
    urllib.request.urlretrieve(url, "downloaded " + imgs[0]["src"])
    print("downloaded ", imgs[0]["src"])
```

是判断这个 URL 页面是否有<div>与<img>，如果有的就获取第一个<div>的文字，且下载第一个<img>的图像。

3.4.3 爬取程序的改进

1．服务器程序

由于 Web 网站是本地的，因此下载图像非常快，而实际应用中 Web 网站在远程服务器中，由于网络原因可能下载会比较慢。模拟这个过程，可以修改服务器程序如下：

```
import flask
import os
import random
import time

app=flask.Flask(__name__)

def getFile(fileName):
    data=b""
    if os.path.exists(fileName):
        fobj=open(fileName,"rb")
        data=fobj.read()
```

```
            fobj.close()
            #随机等待 1-10 秒
            time.sleep(random.randint(1,10))
        return data


    @app.route("/")
    def index():
        return getFile("books.htm")


    @app.route("/<section>")
    def process(section):
        data=""
        if section!="":
            data=getFile(section)
        return data

    if __name__=="__main__":
        app.run()
```

该程序在每次返回一个网页或者图像的函数 getFile()中随机等待了 1～10 秒，这个过程模拟了网络条件较差的情景，访问任何一个网页或者图像都有 1～10 秒的延迟。

2．客户端程序

服务端程序显示在下载图像时需等待一段时间，如果这个图像很大，那么下载时间很长，程序就必须一直等待，无法继续访问其他网页。为了避免这个问题，可以对程序做以下优化：

① 设置 urllib.request 下载图像的时间，如果超过一定时间还没有完成下载就放弃。

② 设置下载过程是一个与主线程不同的子线程，子线程完成下载任务，不影响主线程继续访问别的网页。

改进后客户端程序如下：

```
from bs4 import BeautifulSoup
import urllib.request
import threading


def download(url,fileName):
    try:
        #设置下载时间最长 100 秒
        data=urllib.request.urlopen(url,timeout=100)
        data=data.read()
        fobj=open("downloaded "+fileName,"wb")
        fobj.write(data)
        fobj.close()
```

```python
            print("downloaded ", fileName)
        except Exception as err:
            print(err)


def spider(url):
    global urls
    if url not in urls:
        urls.append(url)
        try:
            data=urllib.request.urlopen(url)
            data=data.read()
            data=data.decode()
            soup=BeautifulSoup(data,"lxml")
            print(soup.find("h3").text)
            links=soup.select("a")
            divs=soup.select("div")
            imgs=soup.select("img")
            if len(divs)>0 and len(imgs)>0:
                note=divs[0].text
                print(note)
                url=start_url+"/"+imgs[0]["src"]

                #启动一个下载线程下载图像
                T=threading.Thread(target=download,args=(url,imgs[0]["src"]))
                T.setDaemon(False)
                T.start()
                threads.append(T)

            for link in links:
                href=link["href"]
                url=start_url+"/"+href
                spider(url)
        except Exception as err:
            print(err)


start_url="http://127.0.0.1:5000"
urls=[]
threads=[]
spider(start_url)
#等待所有线程执行完毕
for t in threads:
```

```
        t.join()
    print("The End")
```

程序运行结果：

```
    计算机
    数据库
    MySQL 数据库

    MySQL 是一个关系数据库管理系统，由瑞典 MySQL AB 公司开发，目前属于 Oracle
旗下产品。MySQL 是最流行的关系数据库管理系统之一，在 Web 应用方面，MySQL 是
最好的 RDBMS（Relational Database Management System，关系数据库管理系统）应用软件。

    计算机
    downloaded   mysql.jpg
    程序设计
    Python 程序设计

    Python （英国发音：/ˈpaɪθən/ 美国发音：/ˈpaɪθɑːn/），是一种面向对象的解释型计
算机程序设计语言，由荷兰人 Guido van Rossum 于 1989 年发明，第一个公开发行版发行
于 1991 年。

    Java 程序设计

    Java 是一门面向对象编程语言，不仅吸收了 C++语言的各种优点，还摒弃了 C++里
难以理解的多继承、指针等概念，因此 Java 语言具有功能强大和简单易用两个特征。Java
语言作为静态面向对象编程语言的代表，极好地实现了面向对象理论，允许程序员以优雅
的思维方式进行复杂的编程。

    downloaded   python.jpg
    计算机网络
    downloaded   java.jpg
    The End
```

可见在访问 python.htm 网页后没有及时完成图片 python.jpg 的下载，python.jpg 是在访问 java.htm 网页后才完成下载的。

3.5 爬取网站的图像文件

3.5.1 项目简介

指定一个网站（例如中国天气网站），可以爬取该网站中的所有图像文件，同时把这些文件保存到程序同级文件夹下 images 子文件夹中。先设计了一个单线程的爬取程序，

项目 3　爬取旅游网站数据

微课 45
爬取网站图像文件

这个程序会因网站的某个图像下载过程缓慢降低整体效率，为了提高爬取的效率还可以设计一个多线程的爬取程序。在多线程程序中如果一个图像下载缓慢，那么仅影响爬取图像的那个线程，不影响其他爬取线程。

3.5.2　单线程爬取图像的程序

```
from bs4 import BeautifulSoup
from bs4 import UnicodeDammit
import urllib.request

def imageSpider(start_url):
    try:
        urls=[]
        req=urllib.request.Request(start_url,headers=headers)
        data=urllib.request.urlopen(req)
        data=data.read()
        dammit=UnicodeDammit(data,["utf-8","gbk"])
        data=dammit.unicode_markup
        soup=BeautifulSoup(data,"lxml")
        images=soup.select("img")
        for image in images:
            try:
                src=image["src"]
                url=urllib.request.urljoin(start_url,src)
                if url not in urls:
                    urls.append(url)
                    print(url)
                    download(url)
            except Exception as err:
                print(err)
    except Exception as err:
        print(err)

def download(url):
    global count
    try:
        count=count+1
        if(url[len(url)-4]=="."):
            ext=url[len(url)-4:]
        else:
            ext=""
        req=urllib.request.Request(url,headers=headers)
```

```
            data=urllib.request.urlopen(req,timeout=100)
            data=data.read()
            fobj=open("images\\"+str(count)+ext,"wb")
            fobj.write(data)
            fobj.close()
            print("downloaded "+str(count)+ext)
    except Exception as err:
        print(err)

start_url="中国天气网/101280601.shtml"
headers = {
    "User-Agent": "Mozilla/5.0 (Windows; U; Windows NT 6.0 x64; en-US; rv:1.9pre) Gecko/2008072421 Minefield/3.0.2pre"}
count=0
imageSpider(start_url)
```

单线程的爬取程序中是逐个下载图像文件的,如果一个文件没有完成下载,后面一个下载任务就必须等待。如果一个文件没有完成下载或者下载中出现问题,将直接影响后续文件的下载,因此效率低,可靠性低。

3.5.3 多线程爬取图像的程序

```
from bs4 import BeautifulSoup
from bs4 import UnicodeDammit
import urllib.request
import threading

def imageSpider(start_url):
    global threads
    global count
    try:
        urls=[]
        req=urllib.request.Request(start_url,headers=headers)
        data=urllib.request.urlopen(req)
        data=data.read()
        dammit=UnicodeDammit(data,["utf-8","gbk"])
        data=dammit.unicode_markup
        soup=BeautifulSoup(data,"lxml")
        images=soup.select("img")
        for image in images:
            try:
```

```
                        src=image["src"]
                        url=urllib.request.urljoin(start_url,src)
                        if url not in urls:
                            print(url)
                            count=count+1
                            T=threading.Thread(target=download,args=(url,count))
                            T.setDaemon(False)
                            T.start()
                            threads.append(T)
                except Exception as err:
                    print(err)
        except Exception as err:
            print(err)


    def download(url,count):
        try:


            if(url[len(url)-4]=="."):
                ext=url[len(url)-4:]
            else:
                ext=""
            req=urllib.request.Request(url,headers=headers)
            data=urllib.request.urlopen(req,timeout=100)
            data=data.read()
            fobj=open("images\\"+str(count)+ext,"wb")
            fobj.write(data)
            fobj.close()
            print("downloaded "+str(count)+ext)
        except Exception as err:
            print(err)


    start_url="/101280601.html"


    headers = {
        "User-Agent": "Mozilla/5.0 (Windows; U; Windows NT 6.0 x64; en-US; rv:1.9pre) Gecko/2008072421 Minefield/3.0.2pre"}
    count=0
    threads=[]


    imageSpider(start_url)
```

```
for t in threads:
    t.join()
print("The End")
```

在这个多线程的爬取程序中下载图像文件是一个线程，因此可以有多个文件在同时下载，而且互不干扰，如果一个文件没有完成下载或者下载中出现问题，也不会影响其他文件的下载，因此效率高，可靠性高。

3.6 爬取图书网站数据

在本书的 1.5 节中创建了一个图书网站，本节将编写一个爬虫程序使用 BeautifulSoup 以及递归方法获取全部页面的数据。

3.6.1 分析网站结构

启动图书网站服务器程序，用浏览器打开 http://127.0.0.1:5000，右击页面在快捷菜单中单击"检查"。可以看到网页的结构代码，如图 3-6-1、图 3-6-2 所示。

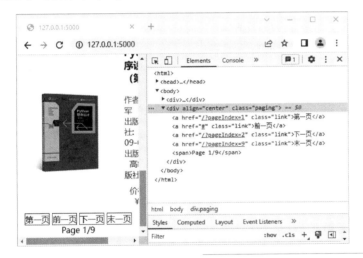

图 3-6-1 换页结构

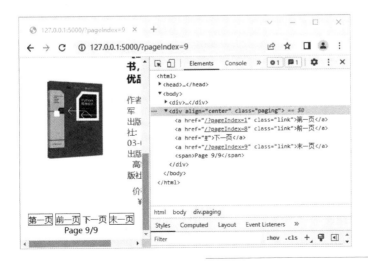

图 3-6-2 最后一页

换页的部分在<div class="paging">中，如果还有下一页就可以找到<a href="...">下一页</a>的元素，如果到了最后一页这个元素变成了<a href="#">下一页</a>，因此通过这个<a>元素就可以获取下一页的地址。

3.6.2 换页递归爬取

先找到<div class="paging">，再找到它下面的第三个<a>元素，它就是下一页的超链接，获取 href 可知下一页地址，方法如下：

```
href=soup.select_one("div[class='paging'] a:nth-child(3)")["href"]
if href!="#":
    nextUrl=urllib.request.urljoin(url,href)
```

在获取下一页的地址 nextUrl 后就可以像爬取第一页那样爬取该页面的数据，因此可以设计一个爬取 URL 页面数据的函数 spider(url)，参数使用 nextUrl 递归调用该函数，完成所有页面的爬取过程，spider()函数结构如下：

```
def spider(url):
    try:
        #爬取 url 页面数据
        #获取下一页地址 nextUrl
        href=soup.select_one("div[class='paging'] a:nth-child(3)")["href"]
        if href!="#":
            nextUrl=urllib.request.urljoin(url,href)
            #递归调用 spider 函数
            spider(nextUrl)
    except Exception as err:
        print(err)
```

3.6.3 图书数据存储

爬取的数据可以存储到一个数据库 books.db 中，该数据库包含一张 books 表，如表 3-6-1 所示。

表 3-6-1 books.db 结构

字段	类型	说明
ID	varchar(8),primary key	编号，关键字
Title	varchar(256)	图书名称
Author	varchar(256)	作者
Publisher	varchar(256)	出版社
PubDate	varchar(16)	出版日期
Price	varchar(16)	价格
Ext	varchar(8)	图书图像扩展名

设计一个数据库管理的文件 database.py，在其中设计一个数据库管理类 Database，这个类包含 initialize()函数，完成数据库表的创建，函数 open()打开数据库，函数 close()关闭数据库，函数 insert()负责插入一条图书记录，show()函数负责显示数据库表数据。

```python
import sqlite3

class Database:
    def open(self):
        self.con=sqlite3.connect("books.db")
        self.cursor=self.con.cursor()

    def close(self):
        self.con.commit()
        self.con.close()

    def initialize(self):
        try:
            self.cursor.execute("drop table books")
            self.con.commit()
        except:
            pass
        sql="""
        create table books (
            ID varchar(8) primary key,
            Title varchar(256),
            Author varchar(256),
            Publisher varchar(256),
            PubDate varchar(16),
            Price varchar(16),
            Ext varchar(8)
        )
        """
        self.cursor.execute(sql)

    def insert(self,ID,Title,Author,Publisher,PubDate,Price,Ext):
        try:
            sql="insert into books (ID,Title,Author,Publisher,PubDate,Price,Ext) values (?,?,?,?,?,?,?)"
            self.cursor.execute(sql,[ID,Title,Author,Publisher,PubDate,Price,Ext])
```

```python
        except:
            pass

    def show(self):
        sql="select ID,Title,Author,Publisher,PubDate,Price,Ext from books"
        self.cursor.execute(sql)
        rows=self.cursor.fetchall()
        for row in rows:
            print(row[0],row[1],row[2],row[3],row[4],row[5],row[6])
        print("Total ",len(rows),"items")
```

3.6.4 编写爬虫程序

根据前文分析，编写爬虫程序如下：

```python
from bs4 import BeautifulSoup
import urllib.request
import os
from database import Database
import threading

def download(src):
    try:
        #获取文件名称
        p=src.rfind("/")
        fn=src[p+1:]
        #读取文件二进制数据
        resp=urllib.request.urlopen(src)
        data=resp.read()
        #保存文件
        f=open("download\\"+fn,"wb")
        f.write(data)
        f.close()
        print("Downloaded",fn)
    except Exception as err:
        print(err)

def spider(url):
    global page,count,threads
    page+=1
```

```python
        print("page",page,url)
        try:
            resp=urllib.request.urlopen(url)
            html=resp.read().decode()
            soup=BeautifulSoup(html,"lxml")
            #获取所有<tr>
            trs=soup.select("table tr")
            for tr in trs:
                count+=1
                ID="%06d"%count
                #查找每本书的数据
                src=tr.select_one("td:first-child img")["src"]
                src=urllib.request.urljoin(url,src)
                p=src.rfind(".")
                Ext=src[p+1:]
                td=tr.select_one("td:last-child")
                Title=td.select_one("div[class='title'] h3").text
                Author=td.select_one("div[class='author'] span:last-child").text
                Publisher=td.select_one("div[class='publisher'] span:last-child").text
                PubDate=td.select_one("div[class='date'] span:last-child").text
                Price=td.select_one("div[class='price'] span:last-child").text
                DB.insert(ID,Title,Author,Publisher,PubDate,Price,Ext)
                #子线程下载图像
                T=threading.Thread(target=download,args=[src])
                T.start()
                #记录线程对象 T
                threads.append(T)
            href=soup.select_one("div[class='paging'] a:nth-child(3)")["href"]
            if href!="#":
                nextUrl=urllib.request.urljoin(url,href)
                spider(nextUrl)
        except Exception as err:
            print(err)


#如果 download 不存在就创建
if not os.path.exists("download"):
    os.mkdir("download")
else:
    #清除 download 的文件
```

```
            fs=os.listdir("download")
            for f in fs:
                os.remove("download\\"+f)
        page=0
        count=0
        threads=[]
        url="http://127.0.0.1:5000"
        DB=Database()
        DB.open()
        DB.initialize()
        spider(url)
        DB.show()
        DB.close()
        #等待全部子线程执行完成
        for T in threads:
            T.join()
        print("Total ",count,"items in",page,"pages")
```

程序启动时先初始化 download 文件夹与数据库 books.db，创建 books 表，之后递归爬取每一页的数据。

注意图像下载是在子线程中完成的，使用 threads 列表记录每个子线程对象，最后使用循环：

```
        for T in threads:
            T.join()
```

该循环等待每个子线程，确保每个子线程都执行完毕后才结束程序。

3.6.5　执行爬虫程序

启动图书网站服务器程序，执行这个爬虫程序爬取了图书网站的记录与图像，books.db 数据库记录，如图 3-6-3 所示。下载的图片，如图 3-6-4 所示。

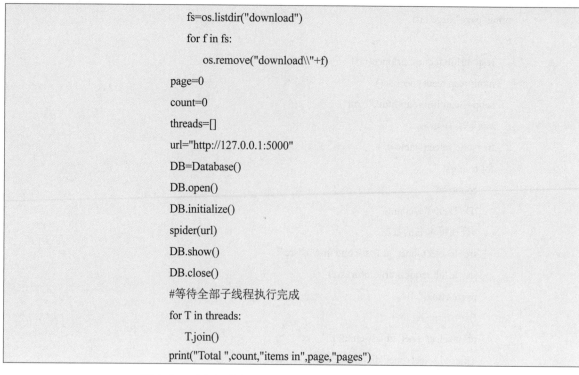

图 3-6-3
数据库数据

图 3-6-4
下载的图片

3.7 实践项目——爬取旅游网站数据

3.7.1 实践项目简介

在中国日报网站有很多旅游景点介绍，进入网站可以看到各个旅游项目，而且每个旅游项目一般都配有精美的图片，通过单击 Next 按钮可进入下一页，如图 3-7-1 所示。

图 3-7-1
中国日报网站

本项目的目的是爬取中国日报网站的 82 个页面中所有旅游项目与对应的图片，把文本数据存储到数据库 travels.db，把图片存储到 download 文件夹。

3.7.2 网站网页分析

要爬取这些页面的文本与图像就必须先分析网页的结构。使用浏览器浏览网站，右击网页，在快捷菜单中选择"检查"，可以看到网页结构，如图 3-7-2 所示。

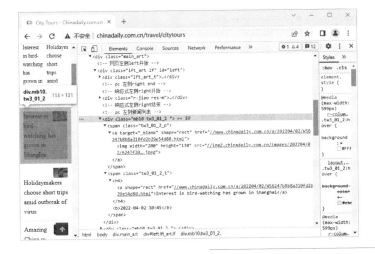

图 3-7-2
网页结构

每个旅游项目都在一个<div>元素中，其中一个结构如下：

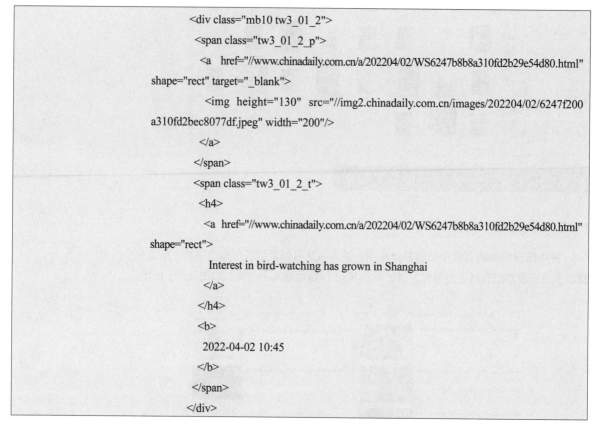

所有的项目都包含在<div class='lft_art lf'>元素中，该元素的下面是 1 个 div[class='mb10 tw3_01_2']元素序列，每个元素是一个旅游项目。

单击其中一个项目，可以看到这个项目的详细内容，如图 3-7-3 所示。详细内容一般包含文字和图像，而且还有很多个页面，本项目只获取第一个页面中的文字内容，包含在一个<div id="Content">元素下面的各个<p>元素中。

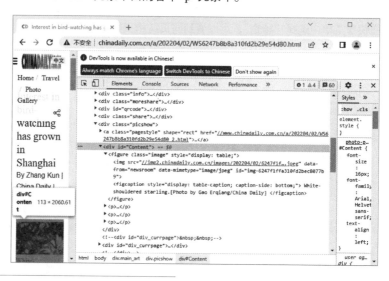

图 3-7-3
详细内容

3.7.3 网站数据爬取

1．获取项目

首先获取<div class='lft art lf'>元素，在该元素中获取所有的<div class="mb10 tw3_01_2">元素，循环遍历获取每个项目：

```
divs=soup.select_one("div[class='lft art lf']").select("div[class=' mb10 tw3_01_2']")
for div in divs:
    #每个 div 是一个项目
```

2．获取标题

显示标题在<h4>元素的<a>中，获取<a>元素中的文本就是标题，编写代码如下：

```
title=div.select_one("h4 a").text
```

3．获取日期

日期在<span>元素下的<b>元素中，编写代码如下：

```
date=div.select_one("span b").text
```

4．获取图像

图像在<span>元素下的<a>元素的<img>中，编写代码如下：

```
src=div.select_one("span a img")["src"]
```

5．获取内容

获取内容的地址，包含在<span>元素下的<h4>的<a>元素中，编写代码如下：

```
href=div.select_one("span h4 a")["href"]
```

在获取地址后再次访问这个 herf 地址获取网页，然后找到<div id="Content">的元素下面的所有<p>元素，这些<p>元素的文本组合在一起就是需要的内容。

3.7.4 网站网页翻页

由于项目分布在很多个页面，在一个页面爬取完毕后要自动进入下一个页面进行爬取，检测翻页的 HTML 代码，如图 3-7-4 所示。

图 3-7-4 网页翻页

翻页按钮 Next 包含在<div class="currpage">元素中，找到该元素下的所有<a class="pagestyle"]元素，再找 Next 文本的<a>元素，那么<a>的 href 就是下页的地址，如果找不到这个地址就表示已经到了最后一页，因此下页地址 nextUrl 可以这样查找：

```
nextUrl=""
links=soup.select("div[id='div_currpage'] a[class='pagestyle']")
for link in links:
    if link.text=="Next":
        href=link["href"]
        if href.startswith("//www."):
            nextUrl="http:"+href
        else:
            nextUrl=urllib.request.urljoin(url,href)
        break
```

3.7.5 网站数据存储

数据可以存储在一个 Sqlite3 数据库 travels.db 中，设计一个数据库表 items，数据包含一个关键字 ID（编号）、tDate（日期）、tTitle（标题）、tContent（内容）以及 tExt（图像扩展名称）。表格结构，如表 3-7-1 所示。

表 3-7-1 items 数据库表

字段名称	字段类型	说明
ID	varchar(6), primary key	序号
tDate	varchar(16)	日期
tTitle	varchar(1024)	标题
tContent	text	内容
tExt	varchar(8)	图像

tExt 只存储图像文件的扩展名，图像文件的名称是 ID 与 tExt 的组合，图像文件被存储在 download 文件夹中。

3.7.6 编写爬虫程序

根据前面的分析，编写完整的爬虫程序如下：

```
from bs4 import BeautifulSoup
import urllib.request
import sqlite3
import os
import time
import threading

class Database:
```

```python
def open(self):
    #打开数据库
    self.con=sqlite3.connect("travels.db")
    self.cursor=self.con.cursor()

def close(self):
    #关闭数据库
    self.con.commit()
    self.con.close()

def initialize(self):
    #初始化数据库,创建 items 表
    try:
        self.cursor.execute("drop table items")
    except:
        pass
    self.cursor.execute("create table items (ID varchar(8) primary key,tDate varchar(16), tTitle varchar(1024),tContent text,tExt varchar(8))")

def insert(self,ID,tDate,tTitle,tContent,tExt):
    #插入一条记录
    try:
        self.cursor.execute("insert into items (ID,tDate,tTitle,tContent,tExt) values (?,?,?,?,?)",[ID,tDate,tTitle,tContent,tExt])
    except Exception as err:
        print(err)

def show(self):
    #显示数据内容
    self.cursor.execute("select ID,tDate,tTitle,tContent,tExt from items order by ID")
    rows=self.cursor.fetchall()
    for row in rows:
        print(row[0])
        print(row[1])
        print(row[2])
        print(row[3])
        print(row[4])
        print()
    print("Total",len(rows),"items")

def downloadImage(ID,src,tExt):
```

```python
            #下载src的图像文件，设置最长下载时间是20秒
            try:
                req=urllib.request.Request(src,headers=headers)
                resp=urllib.request.urlopen(req,timeout=20)
                data = resp.read()
                imgName=ID+"."+tExt
                f=open("download\\"+imgName,"wb")
                f.write(data)
                f.close()
                print("Downloaded "+imgName)
            except Exception as err:
                print(err)


def downloadContent(url):
    #下载项目内容
    content=""
    try:
        req=urllib.request.Request(url,headers=headers)
        resp=urllib.request.urlopen(req)
        html=resp.read().decode()
        soup=BeautifulSoup(html,"lxml")
        ps=soup.select("div[id='Content'] p")
        for p in ps:
            content+=p.text+"\n"
    except Exception as err:
        print(err)
    return content


def initializeDownload():
    #初始化download文件夹
    if not os.path.exists("download"):
        os.mkdir("download")
    fs=os.listdir("download")
    for f in fs:
        os.remove("download\\"+f)

def spider(url):
    #爬取url页面的数据
    global page,count,DB,threads
    page=page+1
    print("Page",page,url)
```

```python
try:
    req=urllib.request.Request(url,headers=headers)
    resp=urllib.request.urlopen(req)
    html=resp.read().decode()
    soup=BeautifulSoup(html,"lxml")
    #获取所有项目的 div
    divs=soup.select("div[class='lft_art lf'] div[class='mb10 tw3_01_2']")
    for div in divs:
        #获取标题
        tTitle=div.select_one("span h4").text
        #获取日期
        tDate=div.select_one("span b").text
        #设置序号
        count=count+1
        ID="%06d"%(count)
        #获取图像文件
        img=div.select_one("span a img")
        src=""
        tExt=""
        if img:
            src=urllib.request.urljoin(url,img["src"])
            p=src.rfind(".")
            if p>=0:
                tExt=src[p+1:]
            #启动一个线程下载图像
            T=threading.Thread(target=downloadImage,args=[ID,src,tExt])
            T.start()
            #把各个线程记录在 threads 列表
            threads.append(T)
        #获取内容链接
        link=div.select_one("span h4 a")["href"]
        link=urllib.request.urljoin(url,link)
        #下载内容
        tContent=downloadContent(link)
        #数据存储到数据库
        DB.insert(ID,tDate,tTitle,tContent,tExt)
    #获取下一页地址 nextUrl
    nextUrl=""
    links=soup.select("div[id='div_currpage'] a[class='pagestyle']")
```

```python
            for link in links:
                if link.text=="Next":
                    #找到下一页地址
                    href=link["href"]
                    if href.startswith("//www."):
                        nextUrl="http:"+href
                    else:
                        nextUrl=urllib.request.urljoin(url,href)
                    break
            if nextUrl:
                #递归调用 spider 函数
                spider(nextUrl)
    except Exception as err:
        print(err)


#http 请求头
headers={"user-agent":"Mozilla/5.0 (Macintosh; Intel Mac OS X 10_8_0) AppleWebKit/537.36 (KHTML, like Gecko) Chrome/32.0.1664.3 Safari/537.36"}


while True:
    print("1.Spider")
    print("2.Show")
    print("3.Exit")
    s=input("Please enter(1,2,3):")
    if s=="1":
        #爬取数据
        #download 文件夹初始化
        initializeDownload()
        threads=[]
        page=0
        count=0
        DB=Database()
        #数据库初始化
        DB.open()
        DB.initialize()
        #从第一页开始爬取
        spider(url="http://www.chinadaily.com.cn/travel/citytours")
        DB.close()
        #线程等待，直到所有图像文件下载完毕
```

```
            for T in threads:
                T.join()
            print("Total %d pages, %d items" %(page,count))
        elif s=="2":
            #显示数据库内容
            DB=Database()
            DB.open()
            DB.show()
            DB.close()
        else:
            #退出程序
            break
```

程序开始时启动一个菜单供用户选择，如果选择 1 就开始进行爬取数据的工作，选择 2 则显示数据库中存储的数据，选择 3 则退出程序。

爬取数据时先初始化 download 文件夹，删除该文件夹中已经存在的所有文件，然后初始化数据库表 items，创建 items 表格，开始爬取与存储数据。函数 spider(url)负责爬取 url 这个页面所有项目的数据，包括日期 tDate 与标题 tTitle。

在获取图像文件的 src 后启动一个线程调用 downloadImage()函数下载图像，图像名称使用 ID 与 tExt 组合，下载的图像存储到 download 文件夹。

在获取内容的地址 link 后使用 downloadContent()函数爬取内容中第一个页面的文本数据，然后存储到 items 表的 tContent 字段。

在爬取结束前调用每个线程的 join()函数等待所有的图像下载线程，等到所有的图像下载完毕完成爬取数据。

3.7.7 执行爬虫程序

执行这个爬虫程序，爬取中国日报网站 82 页的数据，约 1631 个项目，存储在数据库的数据，如图 3-7-5 所示。存储在 download 文件夹的图像文件，如图 3-7-6 所示。

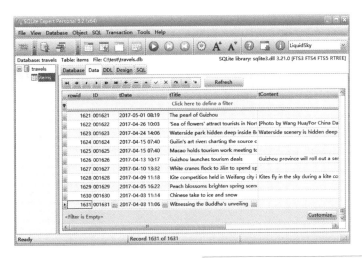

图 3-7-5
数据库数据

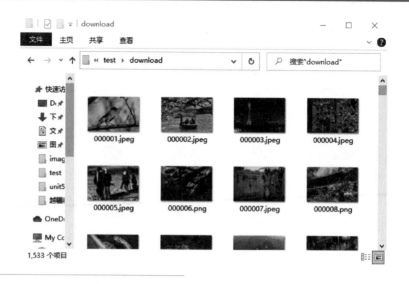

图 3-7-6
图像文件

> 练习三

① 什么是深度优先策略和广度优先策略，它们各有什么特点？

② 分别用深度优先策略与广度优先策略遍历图 3-1-2 中的各个网站，顺序是怎么样的？

③ 一般程序使用函数递归调用去访问各个网站，那么是深度优先还是广度优先策略？

④ 如何启动一个 Python 线程？为什么说爬虫程序一般都会使用多线程？

⑤ 仿照本教程项目 1，设计两个网页 A.html 与 B.html，使其都包含相同结构的学生信息（学号 No、姓名 Name、性别 Gender、年龄 Age），但是两个文件表格中的学生不一样，设计一个包含两个线程的爬虫程序，一个线程爬取 A.html，另外一个线程爬取 B.html，爬取的学生信息都保存到一个共同的列表 students 中，最后显示爬取的结果。

项目4　爬取航空网站数据

本项目将介绍功能强大的 Scrapy 框架爬虫程序的编写方法。Scrapy 是常用的 Python 爬虫框架之一，该框架把数据爬取与数据存储进行了分工，实现了数据的分布式爬取与存储，是一个功能强大的专业爬虫框架。最后介绍爬取与存储航空网站数据的综合案例。

PPT 安装 scrapy 框架

4.1 Scrapy 框架爬虫简介

4.1.1 安装 Scrapy 框架

在 Python 的安装目录中找到 scripts 目录，在 scripts 中执行：

```
pip install scrapy
```

程序执行完毕即完成安装。

微课 46
安装 scrapy 框架

4.1.2 建立 Scrapy 项目

① 进入命令行窗体，在 C 盘中建立一个名为 example 的文件夹，进入 C:\example，执行命令：

```
scrapy startproject demo
```

该命令建立一个名为 demo 的 Scrapy 项目，如图 4-1-1 所示。

微课 47
建立 scrapy 项目

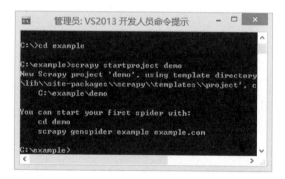

图 4-1-1
建立 Scrapy 项目

② Scrapy 项目建立后会在 C:\example 中建立 demo 文件夹，同时 demo 文件夹下还有另外一个名为 demo 的子文件夹，如图 4-1-2 所示。

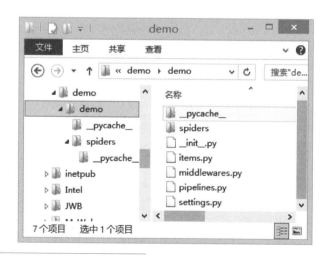

图 4-1-2
Scrapy 的项目结构

③ 使用 PyCharm 打开 example 项目，如图 4-1-3 所示。

4.1 Scrapy 框架爬虫简介

图 4-1-3
使用 PyCharm 打开 example 项目

④ 为测试该 Scrapy 项目，首先要建立一个 Web 网站，在 C:\example 中编写 server.py 程序如下：

```python
import flask

app=flask.Flask(__name__)

@app.route("/")
def index():
    return "测试 Scrapy"

if __name__=="__main__":
    app.run()
```

执行了该程序建立网址为 http://127.0.0.1:5000 的网站，访问该网址显示"测试 Scrapy"。

⑤ 在 C:\example\demo\demo\spider 文件夹中编写名为 MySpider.py 的 Python 爬虫程序如下：

```python
import scrapy

class MySpider(scrapy.Spider):
    name = "mySpider"

    def start_requests(self):
        url ='http://127.0.0.1:5000'
        yield scrapy.Request(url=url,callback=self.parse)

    def parse(self, response):
        print(response.url)
        data=response.body.decode()
        print(data)
```

⑥ 在 C:\example\demo\demo 文件夹中编写名为 run.py 的执行程序如下：

```python
from scrapy import cmdline
```

```
cmdline.execute("scrapy crawl mySpider -s LOG_ENABLED=False".split())
```

⑦ 保存这些程序并运行 run.py，可以在 PyCharm 中看到结果：

```
http://127.0.0.1:5000
测试 Scrapy
```

可见，程序 MySpider.py 访问了 server.py 建立的 Web 网站并爬取了该网站的网页。该项目看起来有点复杂，但是仔细分析也不难理解。MySpider.py 程序分析如下：

```
import scrapy
```

导入 Scrapy 程序包，该包中有一个请求对象 Request 与一个响应对象 Response 类。

```
class MySpider(scrapy.Spider):
    name = "mySpider"
```

任何一个爬虫类都继承于 scrapy.Spider 类，这些类都有一个在爬虫项目中唯一的名字，该爬虫的名字为"mySpider"。

```
def start_requests(self):
    url ='http://127.0.0.1:5000'
    yield scrapy.Request(url=url,callback=self.parse)
```

URL 地址是爬虫程序的入口地址，start_requests()函数是程序的入口函数。程序开始时确定要爬取的网站地址，然后建立一个 scrapy.Request 请求类，向这个类提供要爬取的网页地址。爬取网页完成后就执行默认回调函数 parse()。

📝 注意：

Scrapy 的执行过程是异步进行的，即指定一个网址开始爬取数据时，程序不用一直等待这个网站的响应，Scrapy 提供一个回调函数机制，在爬取网站的同时提供一个回调函数，当网站响应后就执行这个回调函数。

```
def parse(self, response):
    print(response.url)
    data=response.body.decode()
    print(data)
```

回调函数 parse()包含一个 scrapy.Response 类的对象 response，response 对象是网站的一切信息，其中 response.url 是网站的网址，response.body 是网站的二进制数据，即网页的内容。通过 decode()解码变成字符串后可以打印到命令行中显示。

爬虫类就编写完毕，但 MySpider.py 文件中只有一个类，执行后不能爬取数据，需使用 Scrapy 中的命令 scrapy crawl。回到命令行窗体，在 C:\example\demo\demo 中执行命令：

```
scrapy crawl mySpider -s LOG_ENABLED=False
```

就可以看到执行的结果,如图 4-1-4 所示。该命令中 mySpider 是爬虫类的名称,后面的参数是不显示调试信息。

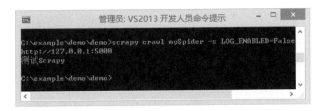

图 4-1-4
命令行执行 Scrapy 程序

但是这种执行方式需要在 PyCharm 与命令行窗体之间频繁切换,为了使用便捷,可以编写一个包含执行命令行语句的 Python 程序 run.py:

```
from scrapy import cmdline
cmdline.execute("scrapy crawl mySpider -s LOG_ENABLED=False".split())
```

在 Pycharm 中运行 run.py 可以执行 MySpider.py 的爬虫程序,与在命令行窗体中执行效果相同,结果直接显示在 PyCharm 中。

4.1.3 入口函数与入口地址

在程序中使用入口函数:

```
def start_requests(self):
    url ='http://127.0.0.1:5000'
    yield scrapy.Request(url=url,callback=self.parse)
```

实际上这个函数也可以用 start_urls() 的入口地址代替:

```
start_urls=['http://127.0.0.1:5000']
```

入口地址可以有多个,因此 start_urls 是一个列表。入口函数和入口地址的作用相同,都是引导函数的开始。

4.1.4 Python 的 yield 语句

入口函数程序如下。可以看到一条 yield 语句,yield 是 Python 的一种特殊语句,作用是返回一个值等待被取走。例如:

```
def fun():
    s=['a','b','c']
    for x in s:
        yield x
    print("fun End")

f=fun()
print(f)
for e in f:
    print(e)
```

程序运行结果：

```
<generator object fun at 0x0000003EA99BD728>
a
b
c
fun End
```

可见，fun()返回一个 generator 对象，这种对象包含一系列的元素，可以使用 for 循环提取，执行循环如下：

```
for e in f:
    print(e)
```

第一次 for e in f 循环，f 执行到 yield 语句，就返回第一个值 a，for 循环从 f 抽取元素 a，执行 e='a'后打印 a。fun()中执行到 yield 时会等待 yield 返回的值被抽走，同时 fun()停留在 yield 语句，a 被抽走后，再次循环，yield 将会返回 b。

第二次 for e in f 循环，抽取 f 中的 b 元素，打印 b，循环继续 yield 将会返回 c 元素。

第三次 for e in f 循环，抽取 f 中的 c 元素，打印 c，循环结束，for e in f 循环中 f 也没有元素可以继续抽取了，显示 fun End，随后程序结束。

包含 yield 语句的函数都会返回一个 generator 的可遍历对象，遍历到 yield 语句时只返回 generator 中一个值，等待调用循环抽取，一旦调用循环抽取后，函数又继续进行。这个过程非常类似两个线程的协作过程，当提供数据的一方准备好数据并使用 yield 数据提交后，等待另外一方把数据抽取，如果未抽取 yield 就一直等待抽取，抽走后数据提供方才继续执行余下程序，等到出现下一次 yield 或者程序结束。

Scrapy 的框架使用执行方式是异步执行，因此大量使用 yield 语句。

> PPT scrapy 中查找 HTML 元素

4.2 Scrapy 中查找 HTML 元素

微课 48
scrapy 中查找 HTML
元素 1

使用 BeautifulSoup 可以查找 HTML 中的元素，Scrapy 中也有同样强大的查找 HTML 元素的功能，那就是使用 XPath 方法。XPath 方法使用 XPath 语法，比 BeautifulSoup 的 select()更灵活而且速度更快。

4.2.1 Scrapy 的 XPath 简介

例 4-2-1：使用 XPath 查找 HTML 中的元素，程序如下：

```
from scrapy.selector import Selector
htmlText=''
<html><body>
<bookstore>
<book>
    <title lang="eng">Harry Potter</title>
    <price>29.99</price>
</book>
```

```
        <book>
          <title lang="eng">Learning XML</title>
          <price>39.95</price>
        </book>
      </bookstore>
    </body></html>
    '''
selector=Selector(text=htmlText)
print(type(selector));
print(selector)
s=selector.xpath("//title")
print(type(s))
print(s)
```

程序运行结果：

class 'scrapy.selector.unified.Selector'>
<Selector xpath=None data='<html><body>\n<bookstore>\n<book>\n <title'>
<class 'scrapy.selector.unified.SelectorList'>
[<Selector xpath='//title' data='<title lang="eng">Harry Potter</title>'>, <Selector xpath='//title' data='<title lang="eng">Learning XML</title>'>]

分析程序的功能：

```
from scrapy.selector import Selector
```

从 Scrapy 中引入 Selector 类，这个类是选择查找类。

```
selector=Selector(text=htmlText)
```

使用 htmlText 的文字建立 Selector 类，装载 HTML 文档，文档装载后形成一个 Selector 对象，就可以使用 XPath 查找元素。

```
print(type(selector))
```

selector 的类型为 scrapy.selector.unified.Selector 对象，这个类型是一个有 XPath 方法的类型。

```
s=selector.xpath("//title")
```

这个方法在文档中查找所有的<title>的元素，其中//表示文档中的任何位置。一般使用 selector.xpath("//tagName")表示在全文档中搜索<tagName>的元素，形成一个 Selector 列表。

```
print(type(s))
```

由于<title>有两个元素，因此这是一个 scrapy.selector.unified.SelectorList 类，类似 scrapy.selector.unified.Selector 的列表。

```
print(s)
```

s 中包含两个 Selector 对象：一个是<Selector xpath='//title' data='<title lang="eng">Harry Potter</title>'>；另一个是<Selector xpath='//title' data='<title lang="eng">Learning XML</title>'>。

可见，Selector 搜索一个<tagName>的 HTML 元素的方法 selector.xpath("//tagName") 在装载 HTML 文档后 selector=Selector(text=htmlText)得到的 selector 是对应全文档顶层元素<html>的，其中//表示全文档搜索，返回结果是一个 Selector 列表，哪怕只有一个元素也会组成一个列表。例如：

- selector.xpath("//body")搜索到 1 个<body>元素，返回结果是 1 个 Selector 列表，列表中包含 1 个 Selector 元素。
- selector.xpath("//title")搜索到 2 个<title>元素，返回结果是 1 个 Selector 列表，列表中包含 2 个 Selector 元素。
- selector.xpath("//book")搜索到 2 个<book>元素，返回结果是 1 个 Selector 列表，列表中包含 2 个 Selector 元素。

微课 50
scrapy 中查找 HTML 元素 2

微课 51
scrapy 中查找 HTML 元素 2

4.2.2 XPath 查找 HTML 元素

① 使用//表示文档下所有节点元素，使用/表示当前节点的下一级节点元素。

例 4-2-2：//与/的使用，程序如下：

```
selector.xpath("//bookstore/book")   #搜索<bookstore>下一级的<book>元素，找到 2 个
selector.xpath("//body/book")        #搜索<body>下一级的<book>元素，结果为空
selector.xpath("//body//book")       #搜索<body>下<book>元素，找到 2 个
selector.xpath("/body//book")        #搜索文档下一级的<body>下的<book>元素，结果为
#空，因为文档的下一级是<html>元素，不是<body>元素
selector.xpath("/html/body//book")   #或者 selector.xpath("/html//book") 搜索<book>
#元素，找到 2 个
selector.xpah("//book/title")        #搜索文档中所有<book>下一级的<title>元素，找到 2
#个，结果与 selector.xpah("//title")、selector.xpath("//bookstore//title")相同
selector.xpath("//book//price")      #与 selector.xpath("//price")结果相同，都是找到 2 个
#<price>元素
```

② 使用.表示当前节点元素，使用 XPath 可以连续调用，如果前一个 XPath 返回一个 Selector 的列表，那么这个列表可以继续调用 XPath，功能是为每个列表元素调用 XPath，最后结果是全部元素调用 XPath 的汇总。

例 4-2-3：使用.进行 XPath 连续调用，程序如下：

```
from scrapy.selector import Selector
htmlText=''
<html>
<body>
<bookstore>
<title>books</title>
<book>
  <title>Novel</title>
    <title lang="eng">Harry Potter</title>
```

```
        <price>29.99</price>
    </book>
    <book>
      <title>TextBook</title>
      <title lang="eng">Learning XML</title>
      <price>39.95</price>
    </book>
</bookstore>
</body></html>
'''
selector=Selector(text=htmlText)
s=selector.xpath("//book").xpath("./title")
for e in s:
    print(e)
```

程序运行结果：

```
<Selector xpath='//book/title' data='<title>Novel</title>'>
<Selector xpath='//book/title' data='<title lang="eng">Harry Potter</title>'>
<Selector xpath='//book/title' data='<title>TextBook</title>'>
<Selector xpath='//book/title' data='<title lang="eng">Learning XML</title>'>
```

可见，使用 selector.xpath("//book")先搜索到文档中所有<book>元素，共有 2 个，然后再次调用 xpath("./title")，从当前<book>元素开始往下一级搜索<title>，每个<book>都可以找到 2 个<title>，因此结果有 4 个<title>。

注意：

如果 XPath 连续调用时，不指定是从前一个 XPath 的结果元素开始，则默认从全文档开始，返回结果不一样。例如：

```
s=selector.xpath("//book").xpath("/title")
```

结果为空，因为 xpath("/title")不是从全文档开始搜索<title>。

```
s=selector.xpath("//book").xpath("//title")
```

返回结果有 10 个元素，因为每个<book>都使用 xpath("//title")在全文档搜索<title>元素，每次都搜索到 5 个<title>元素。

③ 如果 XPath 返回的 Selector 对象列表，再次调用 extract()函数会得到这些对象的元素文本的列表，使用 extract_first()获取列表中第一个元素值，如果列表为空 extract_first()的值为 None。

而对于单一的 Selector 对象，调用 extract()函数可以得到 Selector 对象对应的元素的文本值。单一的 Selector 对象没有 extract_first()函数。

微课 52
scrapy 中查找 HTML
元素 3

例 4-2-4：extract()与 extract_first()函数的使用，程序如下：

```
from scrapy.selector import Selector
htmlText='''
<html>
<body>
<bookstore>
<book id="b1">
    <title lang="english">Harry Potter</title>
    <price>29.99</price>
</book>
<book id="b2">
    <title lang="chinese">学习 XML</title>
    <price>39.95</price>
</book>
</bookstore>
</body></html>
'''
selector=Selector(text=htmlText)
s=selector.xpath("//book/price")
print(type(s),s)
s=selector.xpath("//book/price").extract()
print(type(s),s)
s=selector.xpath("//book/price").extract_first()
print(type(s),s)
```

程序运行结果：

```
<class 'scrapy.selector.unified.SelectorList'> [<Selector xpath='//book/price' data='<price>29.99</price>'>, <Selector xpath='//book/price' data='<price>39.95</price>'>]
<class 'list'> ['<price>29.99</price>', '<price>39.95</price>']
<class 'str'> <price>29.99</price>
```

可见：

- s=selector.xpath("//book/price") 返回的是 SelectorList 列表。
- s=selector.xpath("//book/price").extract() 得到的是由<price>元素的 Selector 对象对应的<price>元素的文本组成的列表，即：

```
['<price>29.99</price>', '<price>39.95</price>']
```

- s=selector.xpath("//book/price").extrac_first() 得到的是由<price>元素的文本组成列表的第一个元素，是一个文本即：

```
<price>29.99</price>
```

④ XPath 使用/@attrName 得到一个 Selector 元素的 attrName 属性节点对象,属性节点对象也是一个 Selector 对象,通过 extract()获取属性值。

例 4-2-5: 获取元素属性值,程序如下:

```
htmlText='''
<html>
<body>
<bookstore>
<book id="b1">
    <title lang="english">Harry Potter</title>
    <price>29.99</price>
</book>
<book id="b2">
    <title lang="chinese">学习 XML</title>
    <price>39.95</price>
</book>
</bookstore>
</body></html>
'''
selector=Selector(text=htmlText)
s=selector.xpath("//book/@id")
print(s)
print(s.extract())
for e in s:
    print(e.extract())
```

程序运行结果:

```
[<Selector xpath='//book/@id' data='b1'>, <Selector xpath='//book/@id' data='b2'>]
['b1', 'b2']
b1
b2
```

可见 s=selector.xpath("//book/@id")的结果是由 2 个<book>的 id 属性组成的 SelectorList 列表,即属性也是一个 Selector 对象。

print(s.extract())的结果是<book>的 id 属性的两个 Selector 对象的属性文本值的列表,即['b1', 'b2']。

```
for e in s:
    print(e.extract())
```

微课 54
scrapy 中查找 HTML
元素 5

列表 s 中的每个 e 都是 Selector 对象,因此 extract()获取对象的属性值。

⑤ XPath 使用/text()得到一个 Selector 元素包含的文本值,文本值节点对象也是一个 Selector 对象,通过 extract()获取文本值。

例 4-2-6：获取节点的文本值，程序如下：

```
from scrapy.selector import Selector
htmlText='''
<html>
<body>
<bookstore>
<book id="b1">
    <title lang="english">Harry Potter</title>
    <price>29.99</price>
</book>
<book id="b2">
    <title lang="chinese">学习 XML</title>
    <price>39.95</price>
</book>
</bookstore>
</body></html>
'''
selector=Selector(text=htmlText)
s=selector.xpath("//book/title/text()")
print(s)
print(s.extract())
for e in s:
    print(e.extract())
```

程序运行结果：

```
[<Selector xpath='//book/title/text()' data='Harry Potter'>, <Selector xpath='//book/title/text()' data='学习 XML'>]
['Harry Potter', '学习 XML']
Harry Potter
学习 XML
```

可见 s=selector.xpath("//book/title/text()")的结果也是 SelectorList 列表，即文本也是一个节点。

print(s.extract())的结果是文本节点的字符串值的列表，即['Harry Potter', '学习 XML']。

```
for e in s:
    print(e.extract())
```

列表 s 中的每个 e 都是 Selector 对象，因此 extract()获取对象的属性值。

注意：
如果一个元素包含的文本不是单一的文本，那么可能会产生多个文本值。

例 4-2-7: 获取多个文本节点值，程序如下：

```python
from scrapy.selector import Selector
htmlText='''
<html>
<body>
<bookstore>
<book id="b1">
    <title lang="english"><b>H</b>arry <b>P</b>otter</title>
    <price>29.99</price>
</book>
</bookstore>
</body></html>
'''
selector=Selector(text=htmlText)
s=selector.xpath("//book/title/text()")
print(s)
print(s.extract())
for e in s:
    print(e.extract())
```

程序运行结果：

```
[<Selector xpath='//book/title/text()' data='arry '>, <Selector xpath='//book/title/text()' data='otter'>]
['arry ', 'otter']
arry
otter
```

可见，<title>中的文本值包含 arry 与 otter 两个。

⑥ XPath 使用 tag[condition]限定一个元素，其中 condition 是由 tag 的属性、文本等计算出的一个逻辑值。如果有多个条件，可以写成：

```
tag[condition1][condition2]...[conditionN]
```

或：

```
tag[condition1 and condition2 and ... and conditionN]
```

微课 55
scrapy 中查找 HTML
元素 6

例 4-2-8: 使用 condition 限定 tag 元素，程序如下：

```python
from scrapy.selector import Selector
htmlText='''
<html>
<body>
<bookstore>
```

```
        <book id="b1">
            <title lang="english">Harry Potter</title>
            <price>29.99</price>
        </book>
        <book id="b2">
            <title lang="chinese">学习 XML</title>
            <price>39.95</price>
        </book>
    </bookstore>
</body></html>
'''
selector=Selector(text=htmlText)
s=selector.xpath("//book/title[@lang='chinese']/text()")
print(s.extract_first())
s=selector.xpath("//book[@id='b1']/title")
print(s.extract_first())
```

程序运行结果：

```
学习 XML
<title lang="english">Harry Potter</title>
```

微课 56
scrapy 中查找 HTML 元素 7

可见，s=selector.xpath("//book/title[@lang='chinese']/text()")搜索<book>下面属性为 lang="chinese"的<title>元素。

s=selector.xpath("//book[@id='b1']/title")搜索<book>下面属性为 id="b1"的<title>元素。

⑦ XPath 可以使用 position()来确定其中一个元素的限制，该选择起始序号是从 1 开始，并非从 0 开始，可以通过 and、or 等构造复杂的表达式。

例 4-2-9：使用 position()序号确定选择的元素，程序如下：

```
from scrapy.selector import Selector
htmlText='''
<html>
<body>
<bookstore>
<book id="b1">
    <title lang="english">Harry Potter</title>
    <price>29.99</price>
</book>
<book id="b2">
    <title lang="chinese">学习 XML</title>
    <price>39.95</price>
</book>
</bookstore>
```

```
</body></html>
"""
selector=Selector(text=htmlText)
s=selector.xpath("//book[position()=1]/title")
print(s.extract_first())
s=selector.xpath("//book[position()=2]/title")
print(s.extract_first())
```

程序运行结果：

```
<title lang="english">Harry Potter</title>
<title lang="chinese">学习 XML</title>
```

其中：

```
s=selector.xpath("//book[position()=1]/title")
s=selector.xpath("//book[position()=2]/title")
```

分别选择第一和第二个<book>元素。

⑧ XPath 使用星号*代表任何元素的节点，不包括 text、comment 的节点。

例 4-2-10： 使用*代表任何元素，程序如下：

```
from scrapy.selector import Selector
htmlText='''
<html>
<body>
<bookstore>
<book id="b1">
  <title lang="english">Harry Potter</title>
  <price>29.99</price>
</book>
<book id="b2">
  <title lang="chinese">学习 XML</title>
  <price>39.95</price>
</book>
</bookstore>
</body></html>
'''
selector=Selector(text=htmlText)
s=selector.xpath("//bookstore/*/title")
print(s.extract())
```

程序运行结果：

```
['<title lang="english">Harry Potter</title>', '<title lang="chinese">学习 XML</title>']
```

其中 s=selector.xpath("//bookstore/*/title")是搜索<bookstore>的孙节点<title>。

⑨ XPath 使用@*代表任何属性

例 4-2-11： 使用@*代表属性，程序如下：

```
from scrapy.selector import Selector
htmlText='''
<html>
<body>
<bookstore>
<book>
  <title lang="english">Harry Potter</title>
  <price>29.99</price>
</book>
<book id="b2">
  <title lang="chinese">学习 XML</title>
  <price>39.95</price>
</book>
</bookstore>
</body></html>
'''
selector=Selector(text=htmlText)
s=selector.xpath("//book[@*]/title")
print(s.extract())
s=selector.xpath("//@*")
print(s.extract())
```

程序运行结果：

```
['<title lang="chinese">学习 XML</title>']
['english', 'b2', 'chinese']
```

其中 s=selector.xpath("//book[@*]/title")是搜索任何包含属性的<book>元素下面的<title>，结果搜索到第二个<book>元素下的<title>。

```
s=selector.xpath("//@*")    #是搜索文档中所有属性节点
```

⑩ XPath 使用 element/parent::*选择元素的父节点，这个节点只有一个。如果写成 element/parent::tag，则为指定 element 名为 tag 的父节点，除非元素的父节点正好为 tag 节点，不然就为 None。

例 4-2-12： XPath 搜索元素的父节点，程序如下：

```
from scrapy.selector import Selector
htmlText='''
<html>
<body>
```

```
<bookstore>
<book>
  <title lang="english">Harry Potter</title>
  <price>29.99</price>
</book>
<book id="b2">
  <title lang="chinese">学习 XML</title>
  <price>39.95</price>
</book>
</bookstore>
</body></html>
'''
selector=Selector(text=htmlText)
s=selector.xpath("//title[@lang='chinese']/parent::*")
print(s.extract())
```

程序运行结果：

```
['<book id="b2">\n   <title lang="chinese">学习 XML</title>\n   <price>39.95</price>\n</book>']
```

其中 s=selector.xpath("//title[@lang='chinese']/parent::*")是查找属性为 lang='chinese'的<title>元素的父节点，也就是 id="b2"的<book>元素。

⑪ XPath 使用 element/folllowing-sibling::*搜索元素后面的同级的所有兄弟节点，使用 element/folllowing-sibling::*[position()=1]搜索元素后面的同级的第一个兄弟节点。

例 4-2-13： 搜索后面的兄弟节点，程序如下：

```
from scrapy.selector import Selector
htmlText="<a>A1</a><b>B1</b><c>C1</c><d>D<e>E</e></d><b>B2</b><c>C2</c>"
selector=Selector(text=htmlText)
s=selector.xpath("//a/following-sibling::*")
print(s.extract())
s=selector.xpath("//a/following-sibling::*[position()=1]")
print(s.extract())
s=selector.xpath("//b[position()=1]/following-sibling::*")
print(s.extract())
s=selector.xpath("//b[position()=1]/following-sibling::*[position()=1]")
print(s.extract())
```

程序运行结果：

```
['<b>B1</b>', '<c>C1</c>', '<d>D<e>E</e></d>', '<b>B2</b>', '<c>C2</c>']
['<b>B1</b>']
['<c>C1</c>', '<d>D<e>E</e></d>', '<b>B2</b>', '<c>C2</c>']
['<c>C1</c>']
```

例如：

```
s=selector.xpath("//b[position()=1]/following-sibling::*[position()=1]")
```

是搜索第一个<b>节点后面的第一个兄弟节点，即<c>C1</c>节点。

⑫ XPath 使用 element/preceding-sibling::* 搜索 element 前面的同级的所有兄弟节点，使用 element/preceding-sibling::*[position()=1] 搜索 element 前面的同级的第一个兄弟节点。

例 4-2-14：搜索前面的兄弟节点，程序如下：

```
from scrapy.selector import Selector
htmlText="<a>A1</a><b>B1</b><c>C1</c><d>D<e>E</e></d><b>B2</b><c>C2</c>"
selector=Selector(text=htmlText)
s=selector.xpath("//a/preceding-sibling::*")
print(s.extract())
s=selector.xpath("//b/preceding-sibling::*[position()=1]")
print(s.extract())
s=selector.xpath("//b[position()=2]/preceding-sibling::*")
print(s.extract())
s=selector.xpath("//b[position()=2]/preceding-sibling::*[position()=1]")
print(s.extract())
```

程序运行结果：

```
[]
['<a>A1</a>', '<d>D<e>E</e></d>']
['<a>A1</a>', '<b>B1</b>', '<c>C1</c>', '<d>D<e>E</e></d>']
['<d>D<e>E</e></d>']
```

例如：

```
s=selector.xpath("//b/preceding-sibling::*[position()=1]")
```

是<b>前面所有兄弟节点中的第一个兄弟节点，因为共有两个<b>节点，因此结果是 ['<a>A1</a>', '<d>D<e>E</e></d>']。

XPath 在 HTML 文档中搜索元素是一种非常实用的技术，限于篇幅这里只讲解了主要的规则，其他具体规则可以查询相关文档，在实践中慢慢熟悉。

4.3 Scrapy 爬取与存储数据

从一个网站爬取数据后往往要将数据存储到数据库中，Scrapy 框架的存储方法十分方便。本节建立一个简单的网站说明这个存储过程，然后编写 Scrapy 爬虫程序爬取数据，最后存储数据。

4.3.1 建立 Web 网站

该网站有一个网页，显示计算机教材，编写 Flask 程序如下：

微课 57
建立 Web 网站 1

```
import flask
app=flask.Flask(__name__)

@app.route("/")
def index():
    html="""
    <books>
    <book>
        <title>Python 程序设计</title>
        <author>小明</author>
        <publisher>高等教育出版社</publisher>
    </book>
    <book>
        <title>Java 程序设计</title>
        <author>小红</author>
        <publisher>高等教育出版社</publisher>
    </book>
    <book>
        <title>MySQL 数据库</title>
        <author>小李</author>
        <publisher>高等教育出版社</publisher>
    </book>
    </books>
    """
    return html

if __name__=="__main__":
    app.run()
```

访问这个网站时返回 XML 的数据，包含教材的名称、作者与出版社信息。

4.3.2 编写数据项目类

该程序要爬取的数据是多本教材，每本教材有教材名称、作者和出版社，因此要建立一个教材类，类中包含 title（教材名称）、author（作者）与 publisher（出版社）。在 C:\example\demo\demo 目录下的 items.py 文件可以用来设计数据项目类，打开该文件，优化文件如下：

微课 58
建立 Web 网站 1

```
import scrapy
class BookItem(scrapy.Item):
    # define the fields for your item here like:
    title = scrapy.Field()
    author=scrapy.Field()
    publisher=scrapy.Field()
```

BookItem 是所设计的教材项目类，这个类必须从 scrapy.Item 类继承，在类中定义教材的字段项目，每个字段项目都是一个 scrapy.Field 对象，这里定义了 3 个字段项目，用来存储 title（教材名称）、author（作者）、publisher（出版社）。

如果 item 是一个 BookItem 对象，那么可以通过 item["title"]、item["author"]、item["publisher"]来获取与设置各个字段的值。例如：

```
item=BookItem()
item["title"]="Python 程序设计"
item["author"]="小明"
item["publisher"]="高等教育出版社"
print(item["title"])
print(item["author"])
print(item["publisher"])
```

4.3.3 编写爬虫程序 mySpider.py

编写爬虫程序如下：

```
import scrapy
from demo.items import BookItem

class MySpider(scrapy.Spider):
    name = "mySpider"
    start_urls=['http://127.0.0.1:5000']

    def parse(self, response):
        try:
            data=response.body.decode()
            selector=scrapy.Selector(text=data)
            books=selector.xpath("//book")
            for book in books:
                item=BookItem()
                item["title"]=book.xpath("./title/text()").extract_first()
                item["author"] = book.xpath("./author/text()").extract_first()
                item["publisher"] = book.xpath("./publisher/text()").extract_first()
                yield item
        except Exception as err:
            print(err)
```

该程序访问 http://127.0.0.1:5000 的网站，返回网页中教材的信息，程序执行过程如下：

```
from demo.items import BookItem
```

从 demo 文件夹的 items.py 文件中引入 BookItem 类的定义。

```
data=response.body.decode()
selector=scrapy.Selector(text=data)
books=selector.xpath("//book")
```

解码网站数据并建立 Selector 对象，搜索所有的<book>节点的元素。

```
for book in books:
    item=BookItem()
    item["title"]=book.xpath("./title/text()").extract_first()
    item["author"] = book.xpath("./author/text()").extract_first()
    item["publisher"] = book.xpath("./publisher/text()").extract_first()
    yield item
```

在每个<book>节点下面搜索到<title>节点，取出文本即教材名称，其中 book.xpath("./title/text()")搜索<book>下的<title>节点中的文本时，代码中一定不能缺少./，该部分表示从当前节点<book>往下搜索。按同样的方法，搜索<author>、<publisher>节点中的文本。使用这些信息组成一个 BookItem 对象，该对象通过语句：yield item 向上一级调用函数返回，接下来 Scrapy 会把这个对象推送给与 items.py 同级目录下的 pipelines.py 文件中的数据管道执行类处理数据。

4.3.4 编写数据管道处理类

在 Scrapy 框架中有的 C:\example\demo\demo 目录的 pipelines.py 文件是数据管道处理类文件，打开该文件可以看到一个默认的管道类，优化数据管道类如下：

```
class BookPipeline(object):
    count=0
    def process_item(self, item, spider):
        BookPipeline.count+=1
        try:
            if BookPipeline.count==1:
                fobj=open("books.txt","wt")
            else:
                fobj=open("books.txt","at")
            print(item["title"], item["author"], item["publisher"])
            fobj.write(item["title"]+","+item["author"]+","+item["publisher"]+"\n")
            fobj.close()
        except Exception as err:
            print(err)
        return item
```

该类命名为 BookPipeline，继承自 object 类，此类中最重要的函数是 process_item()，Scrapy 爬取数据开始时会建立一个 BookPipeline 类对象，然后每爬取一个数据类 BookItem 项目 item，mySpider 程序会把这个对象推送给 BookPipeline 对象，同时调用 process_item()

函数一次。process_item()函数参数中的 item 就是推送的数据，于是就可以在这个函数中保存爬取的数据了。

> **注意：**
> Scrapy 要求 process_item()函数最后返回 item 变量。

该程序中使用文件存储爬取的数据，BookPipeline 类中先定义一个类属性 count，用来记录 process_item()被调用的次数。如果是第一次调用（count 为 1），那么就执行语句 fobj=open("books.txt","wt")新建立一个 books.txt 文件，然后把 item 的数据写到该文件中。如果不是第一次调用（count 大于 1 时），就执行语句 fobj=open("books.txt","at")打开已经存在的文件 books.txt，把 item 的数据追加到文件中。如此在反复执行爬虫程序的过程时保证每次清除掉上次的数据，记录本次爬取的数据。

4.3.5 设置 Scrapy 的配置文件

mySpider 程序执行后每爬取一个 item 项目都会推送到 BookPipelines 类并调用 process_item()函数，那么 Scrapy 怎样执行这个流程的前提是必须设置这样一个通道。

在 demo 文件夹下有一个 settings.py 的设置文件，打开该文件可以看到很多设置项目，大部分是注释语句，找到语句 ITEM_PIPELINES，设置成如下形式：

```
# Configure item pipelines
# See http://scrapy.readthedocs.org/en/latest/topics/item-pipeline.html

ITEM_PIPELINES = {
    'demo.pipelines.BookPipeline': 300,
}
```

ITEM_PIPELINES 是一个字典，把关键字改成 demo.pipelines.BookPipeline'，而 BookPipelines 就是在 pipelines.py 文件中设计的数据管道类的名称，后面的 300 是一个默认的整数。

设置后就连通了爬虫程序 mySpider 和数据管道处理程序 pipelines.py 的通道，Scrapy 工作时会把 mySpider 爬虫程序通过 yield 返回的每项数据推送给 pipelines.py 程序的 BookPipeline 类，并执行 process_item()函数，这样就可以保存数据了。

从上面的分析可以看到 Scrapy 把数据爬取与数据存储分开处理，且异步执行，mySpider 每爬取到一个数据项目 item，就使用 yield 推送给 pipelines.py 程序存储，等待存储完毕后又再次爬取另外一个数据项目 item，再次使用 yield 推送到 pipelines.py 程序，然后再次存储，这个过程一直进行到爬取过程结束，文件 books.txt 中就存储了所有的爬取数据了。

4.4 Scrapy 爬取网站数据

为了模拟使用 Scrapy 爬取网站多个网页数据的过程，首先使用 Flask 搭建一个小型的 Web 网站。

4.4.1 建立 Web 网站

（1）books.htm

```
<h3>计算机</h3>
<ul>
<li><a href="database.htm">数据库</a></li>
<li><a href="program.htm">程序设计</a></li>
<li><a href="network.htm">计算机网络</a></li>
</ul>
```

（2）database.htm

```
<h3>数据库</h3>
<ul>
<li><a href="mysql.htm">MySQL 数据库</a></li>
</ul>
<a href="books.htm">Home</a>
```

（3）program.htm

```
<h3>程序设计</h3>
<ul>
<li><a href="python.htm">Python 程序设计</a></li>
<li><a href="java.htm">Java 程序设计</a></li>
</ul>
<a href="books.htm">Home</a>
```

（4）network.htm

```
<h3>计算机网络</h3>
<a href="books.htm">Home</a>
```

（5）mysql.htm

```
<h3>MySQL 数据库</h3>
<a href="books.htm">Home</a>
```

（6）python.htm

```
<h3>Python 程序设计</h3>
<a href="books.htm">Home</a>
```

（7）java.htm

```
<h3>Java 程序设计</h3>
<a href="books.htm">Home</a>
```

需要编写一个爬虫程序爬取这个网站所有页面的<h3>标题文字。首先编写服务器程序如下：

```
import flask
import os
app=flask.Flask(__name__)

def getFile(fileName):
    data=b""
    if os.path.exists(fileName):
        fobj=open(fileName,"rb")
        data=fobj.read()
        fobj.close()
    return data

@app.route("/")
def index():
    return getFile("books.htm")

@app.route("/<section>")
def process(section):
    data=""
    if section!="":
        data=getFile(section)
    return data

if __name__=="__main__":
    app.run()
```

4.4.2 编写 Scrapy 爬虫程序

微课 60
建立 Web 网站 2

重新编写 mySpider.py 程序如下：

```
import scrapy

class MySpider(scrapy.Spider):
    name = "mySpider"
    start_urls=['http://127.0.0.1:5000']

    def parse(self, response):
        try:
            print(response.url)
            data=response.body.decode()
```

```
                selector=scrapy.Selector(text=data)
                print(selector.xpath("//h3/text()").extract_first())
                links=selector.xpath("//a/@href").extract()
                for link in links:
                    url=response.urljoin(link)
                    yield scrapy.Request(url=url,callback=self.parse)
        except Exception as err:
            print(err)
```

运行 run.py 的结果：

```
http://127.0.0.1:5000
计算机
http://127.0.0.1:5000/network.htm
计算机网络
http://127.0.0.1:5000/program.htm
程序设计
http://127.0.0.1:5000/database.htm
数据库
http://127.0.0.1:5000/java.htm
Java 程序设计
http://127.0.0.1:5000/python.htm
Python 程序设计
http://127.0.0.1:5000/books.htm
计算机
http://127.0.0.1:5000/mysql.htm
MySQL 数据库
```

Scrapy 自动筛选已经访问过的网页，程序的执行过程分析如下：

```
start_urls=['http://127.0.0.1:5000']
```

start_urls 是入口地址，成功访问这个地址后会回调 parse()函数。

```
def parse(self, response):
```

回调函数的 response 对象包含了网站返回的信息。

```
data=response.body.decode()
selector=scrapy.Selector(text=data)
```

网站返回的 response.body 是二进制数据，要使用 decode()转为文本，然后建立 Selector 对象。

```
print(selector.xpath("//h3/text()").extract_first())
```

获取网页中的<h3>标题中的文本，该文本就是要爬取的数据，这个数据只有一项。

```
links=selector.xpath("//a/@href").extract()
```

获取所有的<a href=...>链接的 href 值，组成 links 列表。

```
for link in links:
    url=response.urljoin(link)
    yield scrapy.Request(url=url,callback=self.parse)
```

遍历 links 的每个 link，通过 urljoin()函数与 response.url 地址组合成完整的 url 地址，再次建立 Request 对象，回调函数仍然为 parse()，即这个 parse()函数会被递归调用。其中使用了 yield 语句返回每个 Request 对象，这是 Scrapy 程序的规则。

4.4.3 存储 Scrapy 爬取的数据

爬虫程序爬取的数据要通过管道文件存储。管道文件 pipelines.py 中包含一个 BookPipeline 的类，这个类有 open_spider()与 close_spider()两个重要函数。

（1）open_spider()函数

Scrapy 爬虫程序开始时会创建一个 BookPipeline 的对象，并调用 open_spider()函数，可以在这个函数中准备数据存储，例如打开文件或数据库等。

（2）close_spider()函数

Scrapy 爬虫程序在结束时会自动调用 close_spider()函数，可以在这个函数中保存数据之后的善后工作，例如关闭文件或数据库。

根据这个原则，编写 BookPipeline 类如下：

```python
class BookPipeline(object):
    def open_spider(self,spider):
        print("opened")
        self.fobj=open("books.txt","wt")
        self.opened=True

    def close_spider(self, spider):
        print("closed")
        if self.opened:
            self.fobj.close()

    def process_item(self, item, spider):
        try:
            self.fobj.write(item["title"]+"\n")
        except Exception as err:
            print(err)
        return item
```

可见，在 open_spider()中打开了 books.txt 文件，使用 self.fobj 记录文件的对象，在 process_item()中把每个 item 数据写入文件，在 close_spider()中关闭文件。

重新修改 MySpider.py 爬虫程序,在爬取数据后 yield 返回数据给管道程序进行存储,MySpider.py 如下:

```python
import scrapy
from demo.items import BookItem

class MySpider(scrapy.Spider):
    name = "mySpider"
    start_urls=['http://127.0.0.1:5000']

    def parse(self, response):
        try:
            print(response.url)
            data=response.body.decode()
            selector=scrapy.Selector(text=data)
            title=selector.xpath("//h3/text()").extract_first()
            print(title)
            item=BookItem()
            item["title"]=title
            yield item
            links=selector.xpath("//a/@href").extract()
            for link in links:
                url=response.urljoin(link)
                yield scrapy.Request(url=url,callback=self.parse)
        except Exception as err:
            print(err)
```

运行该爬虫程序,可以看到 books.txt 中存储了爬取的数据:

```
计算机
计算机网络
程序设计
数据库
计算机
MySQL 数据库
Java 程序设计
Python 程序设计
```

4.5 实践项目——爬取图书网站数据

本书 1.5 节中设计了一个图书网站,本节使用 Scrapy 设计一个爬虫程序爬取该网站的所有数据与图像。

4.5.1 网站结构分析

网站结构分析详见 3.6 节。

4.5.2 图书数据爬取

微课 62
网站图书数据分析 2

（1）创建启动函数

创建一个名称为 demo 的 Scrapy 项目，编写 demo/spiders/MySpider.py 文件中 MySpider 类的启动函数，程序如下：

```
class MySpider(scrapy.Spider):
    name = "mySpider"

    def start_requests(self):
        url='http://127.0.0.1:5000/'
        yield scrapy.Request(url=url,callback=self.parse)
```

其中 url 是入口地址，函数生成一个 Request 请求，并回调函数 parse()。

（2）获取各个项目

图书的各个项目包含在 <table> 的 <tr> 元素中，循环遍历每个 <tr> 元素，可以找到每本书的图像与数据：

```
selector=scrapy.Selector(text=response.body.decode())
trs=selector.xpath("//table//tr")
for tr in trs:
    #查找每本书的数据
```

（3）获取图书图像

<tr> 中的第一个 <td> 就包含图像元素 <img>，获取图像 src：

```
td=tr.xpath("./td[position()=1]")[0]
src=td.xpath(".//img/@src").extract_first()
src=response.urljoin(src)
request=scrapy.Request(url=src,callback=self.download)
```

其中 urljoin() 函数把 url 与 src 组合成完整的地址。在图像地址 src 获取后可以设计下载图像的函数，把图像下载到 download 文件夹，函数如下：

```
def download(self,response):
    fileName = response.meta["fileName"]
    try:
        f=open("download\\"+fileName,"wb")
        f.write(response.body)
        f.close()
        print("downloaded",fileName)
```

```
except Exception as err:
    print(fileName,str(err))
```

（4）获取图书数据

在<tr>元素的最后一个<td>中包含了图书数据，分析网页结构发现图书数据包括 title（图书的名称）、author（作者）、publisher（出版社）、pubDate（出版日期）、price（价格）等，获取方法如下：

```
td=tr.xpath("./td[position()=2]")[0]
Title=td.xpath(".//div[@class='title']//h3/text()").extract_first()
Author=td.xpath(".//div[@class='author']//span[@class='attrs']/text()").extract_first()
Publisher=td.xpath(".//div[@class='publisher']//span[@class='attrs']/text()").extract_first()
PubDate=td.xpath(".//div[@class='date']//span[@class='attrs']/text()").extract_first()
Price = tr.xpath(".//div[@class='price']/span[@class='price']/text()").extract_first()
```

（5）获取换页地址

网页换页先找到<div class="paging">，再找到该元素下面的第三个<a>元素，它就是下一页的超链接，获取 href 可知下一页地址，方法如下：

```
#最后一页时，link 为#
link=selector.xpath(".//div[@class='paging']/a[position()=3]/@href").extract_first()
if link!="#":
    url=response.urljoin(link)
    yield scrapy.Request(url=url, callback=self.parse)
```

在获取下一页的地址 url 后就生成一个 Request 请求，回调函数是 parse()，就可以像爬取第一页那样爬取这个页面的数据。

4.5.3 图书数据存储

爬取的数据可以存储到数据库 books.db 中。该数据库包含一张 books 表，结构如表 3-6-1 所示。

在 pipelines.py 文件中编写 BookPipeline 类，该类主体结构如下：

```
class BookPipeline(object):
    def open_spider(self,spider):
        #爬虫程序开始时执行
        #数据库初始化与 download 文件夹初始化

    def close_spider(self,spider):
        #爬虫程序结束时执行
        #关闭数据库

    def insertDB(self,ID,Title,Author,Publisher,PubDate,Price,Ext):
        #插入数据库记录
```

项目 4　爬取航空网站数据

```
def process_item(self, item, spider):
    #爬取到一个项目 item 时执行
    #调用 insertDB 函数插入数据
    return item
```

该类包含 4 个函数，其中 open_spider()与 close_spider()函数是在爬虫程序开始和结束时执行的函数，在这两个函数中完成数据库初始化与关闭工作。process_item()函数是一个主要函数，它在 MySpider 中的 yield item 语句触发时执行，每次爬取到一个项目就执行一次这个函数，item 就是一个 BookItem 对象，在这个函数中调用 insertDB()函数把数据插入数据库中保存。

4.5.4　设计爬虫程序

（1）编写 items.py

根据数据库表中各字段的要求，设计 items.py 文件的 BookItem 类如下：

```python
import scrapy

class BookItem(scrapy.Item):
    # define the fields for your item here like:
    ID=scrapy.Field()
    Title = scrapy.Field()
    Author = scrapy.Field()
    Publisher = scrapy.Field()
    PubDate = scrapy.Field()
    Price = scrapy.Field()
    Ext = scrapy.Field()
```

（2）编写 MySpider.py

根据前面的分析，MySpiper.py 文件主要包含 MySpider 类，程序如下：

```python
import scrapy
from demo.items import BookItem

class MySpider(scrapy.Spider):
    name = "mySpider"

    def start_requests(self):
        self.count=0
        url='http://127.0.0.1:5000/'
        yield scrapy.Request(url=url,callback=self.parse)

    def parse(self, response):
        try:
            selector=scrapy.Selector(text=response.body.decode())
```

```
trs=selector.xpath("//table//tr")
for tr in trs:
    self.count+=1
    ID="%06d" %self.count
    td=tr.xpath("./td[position()=2]")[0]
    Title=td.xpath(".//div[@class='title']//h3/text()").extract_first()
    Author=td.xpath(".//div[@class='author']//span[@class='attrs']/text()").extract_first()
    Publisher=td.xpath(".//div[@class='publisher']//span[@class='attrs']/text()").extract_first()
    PubDate=td.xpath(".//div[@class='date']//span[@class='attrs']/text()").extract_first()
    Price = tr.xpath(".//div[@class='price']/span[@class='price']/text()").extract_first()
    td=tr.xpath("./td[position()=1]")[0]
    src=td.xpath(".//img/@src").extract_first()
    Ext=""
    if src:
        #下载图像
        p=src.rfind(".")
        Ext=""
        if p>=0:
            Ext=src[p:]
        src=response.urljoin(src)
        request=scrapy.Request(url=src,callback=self.download)
        request.meta["fileName"]=ID+Ext
        yield request
    #如果某个数据字段不存在，则 extract_first()返回 None，并把数据设置为空字符串
    item=BookItem()
    item["ID"]=ID
    item["Title"]=Title.strip() if Title else ""
    item["Author"]=Author.strip() if Author else ""
    item["PubDate"] = PubDate.strip() if PubDate else ""
    item["Publisher"] = Publisher.strip() if Publisher else ""
    item["Price"] = Price.strip() if Price else ""
    item["Ext"]=Ext
    yield item
    #最后一页时，link 为#
```

```
                            link=selector.xpath("//div[@class='paging']/a[position()=3]/@href").
            extract_first()
                            if link!="#":
                                url=response.urljoin(link)
                                yield scrapy.Request(url=url, callback=self.parse)
                    except Exception as err:
                        print(err)


            def download(self,response):
                fileName = response.meta["fileName"]
                try:
                    f=open("download\\"+fileName,"wb")
                    f.write(response.body)
                    f.close()
                    print("downloaded",fileName)
                except Exception as err:
                    print(fileName,str(err))
```

在这个类中设计一个 self.count 变量，记录项目的数目。

（3）编写 pipelines.py 文件

在 pipelines.py 文件中主要包含 BookPipeline 类，程序如下：

```
import sqlite3
import os

class BookPipeline(object):
    def open_spider(self,spider):
        print("open_spider")
        self.con=sqlite3.connect("books.db")
        self.cursor=self.con.cursor()
        try:
            self.cursor.execute("drop table books")
        except:
            pass
        sql="""
        create table books (
            ID varchar(8) primary key,
            Title varchar(256),
            Author varchar(256),
            Publisher varchar(256),
            PubDate varchar(256),
            Price varchar(256),
```

```
                    Ext varchar(8))
                    """
                    self.cursor.execute(sql)
                    if not os.path.exists("download"):
                        os.mkdir("download")
                    else:
                        fs=os.listdir("download")
                        for f in fs:
                            os.remove("download\\"+f)

            def close_spider(self,spider):
                print("close_spider")
                print("Total ",spider.count)
                self.con.commit()
                self.con.close()

            def insertDB(self,ID,Title,Author,Publisher,PubDate,Price,Ext):
                try:
                    sql="insert into books (ID,Title,Author,Publisher,PubDate,Price,Ext) values (?,?,?,?,?,?,?)"
                    self.cursor.execute(sql,[ID,Title,Author,Publisher,PubDate,Price,Ext])
                except Exception as err:
                    print(err)

            def process_item(self, item, spider):
                ID=item["ID"]
                Title=item["Title"]
                Author=item["Author"]
                Publisher=item["Publisher"]
                PubDate=item["PubDate"]
                Price=item["Price"]
                Ext=item["Ext"]
                print(ID,Title)
                self.insertDB(ID,ID,Title,Author,Publisher,PubDate,Price,Ext)
                return item
```

（4）配置 settings.py 文件

在配置文件中设置管道为 pipelines.py 中的 BookPipeline 类：

```
# Configure item pipelines
# See https://docs.scrapy.org/en/latest/topics/item-pipeline.html
ITEM_PIPELINES = {
```

```
                    'demo.pipelines.BookPipeline': 300,
                }
```

4.5.5 执行爬虫程序

编写 run.py 文件：

```
from scrapy import cmdline
cmdline.execute("Scrapy crawl mySpider -s LOG_ENABLED=False".split())
```

执行 run.py 文件，程序运行后爬取到了图书网站的全部图书数据与图片，结果与 3.6 节相同。

4.6 实践项目——爬取航空网站数据

4.6.1 项目简介

航空网站有国内外主要机场的航班进出港数据，浏览该网站，输入城市（如深圳）名称，再单击"进港"就可以看到机场进港的各个航班，如图 4-6-1 所示。

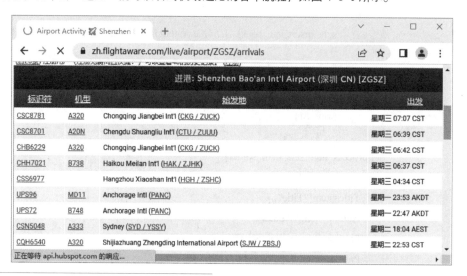

图 4-6-1 进港航班

除了进港航班数据，还有出港航班数据，结构与进港的很相似。

每条数据包括航班号、机型、出发地机场名称、出发时间、到达时间等。这样的数据非常多，每个页面显示 20 条数据，单击"后 20 条"可以看到下一个页面的后 20 条数据。

4.6.2 网页结构分析

右击进港航班数据，在快捷菜单中选择"检查"，看到网页结构，如图 4-6-2 所示。所有的航班数据都包含在一个<table class="prettyTable fullWidth">的表中，表的<tbody>下各<tr>元素包含的就是各个航班的数据。

4.6 实践项目——爬取航空网站数据

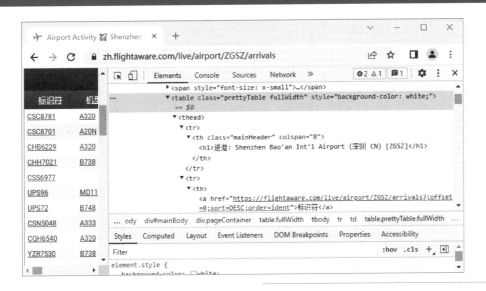

图 4-6-2
网页结构

复制<table class="prettyTable fullWidth">元素代码，使用 BeautifulSoup 的 pretty()函数整理，主体结构如下：

```
<table class="prettyTable fullWidth" style="background-color: white;">
    <thead>
     <tr>
      <th class="mainHeader" colspan="8">
       <h1>
        进港: Shenzhen Bao'an Int'l Airport (深圳 CN) [ZGSZ]
       </h1>
      </th>
     </tr>
     <tr>
      <th>
       <a href="https://flightaware.com/live/airport/ZGSZ/arrivals?;offset=0;sort=DESC;order=ident">
        标识符
       </a>
      </th>
      <th>
       <a href="https://flightaware.com/live/airport/ZGSZ/arrivals?;offset=0;sort=DESC;order=aircrafttype">
        机型
       </a>
      </th>
      <th>
       <a href="https://flightaware.com/live/airport/ZGSZ/arrivals?;offset=0;sort=DESC;order=origin">
```

```html
            始发地
          </a>
        </th>
        <th>
          <a href="https://flightaware.com/live/airport/ZGSZ/arrivals?;offset=0;sort=DESC;order=actualdeparturetime">
            出发
          </a>
        </th>
        <th>
          <a href="https://flightaware.com/live/airport/ZGSZ/arrivals?;offset=0;sort=DESC;order=actualarrivaltime">
            到达
          </a>
          <a href="https://flightaware.com/live/airport/ZGSZ/arrivals?; offset=0; order=actualarrivaltime;sort=ASC">
              <img height="12" src="https://e0.flightcdn.com/images/live/arrow_down.gif" width="12"/>
          </a>
        </th>
      </tr>
    </thead>
    <tbody>
      <tr>
        <td class="smallrow1" style="text-align: left">
         <span title="四川航空公司">
          <a href="/live/flight/id/CSC8781-1649028000-schedule-0775%3a0">
            CSC8781
          </a>
         </span>
        </td>
        <td class="smallrow1" style="text-align: left">
         <span title="Airbus A320 (双发)">
          <a href="/live/aircrafttype/A320">
            A320
          </a>
         </span>
        </td>
        <td class="smallrow1" style="text-align: left">
         <span class="hint" itemprop="name" original-title="Chongqing Jiangbei Int'l (重庆重庆江北 CN) - CKG / ZUCK">
```

```html
      <span dir="ltr">
        Chongqing Jiangbei Int'l
      </span>
    </span>
    <span dir="ltr">
     (
     <a href="/live/airport/ZUCK" itemprop="url">
       CKG / ZUCK
     </a>
     )
    </span>
   </td>
   <td class="smallrow1" style="text-align: left">
    星期三 07:07
    <span class="tz">
     CST
    </span>
   </td>
   <td class="smallrow1" style="text-align: left">
    星期三 08:46
    <span class="tz">
     CST
    </span>
   </td>
  </tr>
  <tr>
   <td class="smallrow2" style="text-align: left">
    <span title="四川航空公司">
      <a href="/live/flight/id/CSC8701-1649025120-schedule-0653%3a0">
        CSC8701
      </a>
    </span>
   </td>
   <td class="smallrow2" style="text-align: left">
    <span title="Airbus A320neo (双发)">
      <a href="/live/aircrafttype/A20N">
        A20N
      </a>
    </span>
   </td>
   <td class="smallrow2" style="text-align: left">
```

```html
            <span class="hint" itemprop="name" original-title="Chengdu Shuangliu Int'l
(成都 CN) - CTU / ZUUU">
              <span dir="ltr">
                Chengdu Shuangliu Int'l
              </span>
            </span>
            <span dir="ltr">
              (
              <a href="/live/airport/ZUUU" itemprop="url">
                CTU / ZUUU
              </a>
              )
            </span>
          </td>
          <td class="smallrow2" style="text-align: left">
            星期三 06:39
            <span class="tz">
              CST
            </span>
          </td>
          <td class="smallrow2" style="text-align: left">
            星期三 08:38
            <span class="tz">
              CST
            </span>
          </td>
        </tr>
......
    </table>
```

4.6.3 爬取航班数据

可以看到数据都包含在<table class="prettyTable fullWidth">中，因此要先获取这个表的各个<tr>元素，设置 response 是 Scrapy 的 parse()函数中的响应对象，那么如下方法获取这个<table>中的各个<tr>元素：

```
html=response.body.decode()
selector=scrapy.Selector(text=html)
trs=selector.xpath("//table[contains(@class,'prettyTable') and contains(@class,'fullWidth')]//tr")
for tr in trs[2:]:
    #各个航班的数据
    tds=tr.xpath(".//td")
```

4.6 实践项目——爬取航空网站数据

> **注意：**
> 因为<table>的类包含两个值，中间用空格隔开了，因此使用 Xpath 的 contains()函数进行判定。前面两个<tr>行的数据只是标题，真正的数据在 trs[2:]中，各个航班的数据就包含在不同的<td>元素中。

航班号 flight 以及航空公司名称 company 包含在第一个<td>元素的<span>中，从代码结构中爬取方式如下：

```
flight=tds[0].xpath("./span/a/text()").extract_first()
company=tds[0].xpath("./span/@title").extract_first()
```

机型 flightType 包含在第二个<td>的<span>的 title 属性中，因此：

```
flightType=tds[1].xpath("./span/@title").extract_first()
```

出发机场 airport 包含在第三个<td>元素中下面的第一个<span dir="ltr">中，因此：

```
airport=tds[2].xpath(".//span[@dir='ltr'][position()=1]/text()").extract_first()
```

出发时间 departureTime 和到达时间 arrivalTime 分别包含在第 4 和第 5 个<td>中，因此：

```
departureTime=tds[3].xpath("./text()").extract_first()
arrivalTime=tds[4].xpath("./text()").extract_first()
```

4.6.4 获取换页地址

从网站的结构中显示一个页面大约只有 20 条数据，检查"后 20 条"的网页结构，如图 4-6-3 所示。

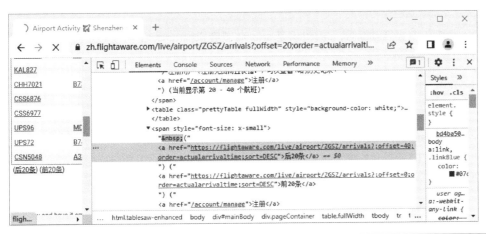

图 4-6-3 换页结构

从网页结构看到在数据表<table class="prettyTable fullWidth">的后面有一个<span>元素，<span>中包含各个<a>超链接，找到一个<a>后 20 条</a>的链接就可以获取换页地址，如果找不到这样的链接说明已经是最后一页，没有后面 20 条数据了。因此，如下的程序可以找到换页地址：

```
                url=""
                links= selector.xpath("//table[contains(@class,'prettyTable') and contains(@class,'fullWidth')]/
                following-sibling::span//a")
                for link in links:
                    s=link.xpath("./text()").extract_first()
                    if s=="后 20 条":
                        href=link.xpath("./@href").extract_first()
                        url=response.urljoin(href)
                        break
```

4.6.5 存储航班数据

爬取的航班数据包括 direction（方向）、flight（航班号）、company（航空公司）、flightType
（机型）、airport（出发或者到达机场）、departureTime（出发时间）、arrivalTime（到达时间），
数据存储在 flights.db 的数据库中，在这个数据库中创建一个 flights 表，SQL 语句如下：

```
create table flights (
    ID varchar(8) primary key,
    direction varchar(32),
    flight varchar(32),
    company varchar(256),
    flightType varchar(64),
    airport varchar(256),
    departureTime varchar(32),
    arrivalTime varchar(32)
)
```

其中 ID 是自动编号，direction 是 arrivals 或者 departures 之一，表示这个航班是进港
航班或者是出港航班。

4.6.6 编写爬虫程序

使用 4.5 节中的 Scrapy 项目结构，在此基础上编写各个文件。

（1）编写 items.py

设置 ID 是项目的编号，在 items.py 中编写一个 FlightItem 类如下：

```
import scrapy
class FlightItem(scrapy.Item):
    # define the fields for your item here like:
    ID=scrapy.Field()
    flight = scrapy.Field()
    direction=scrapy.Field()
    flightType = scrapy.Field()
    company = scrapy.Field()
```

```
            airport = scrapy.Field()
            departureTime = scrapy.Field()
            arrivalTime = scrapy.Field()
```

（2）编写 pipelines.py

在 pipelines.py 中创建一个 FlightPipeline 类，在类的 open_sipder() 函数中初始化数据库，在 close_spider() 函数中关闭数据库，在 process_item() 函数中插入一条航班数据到数据库中。

```
        import sqlite3
        import os
        import datetime

        class FlightPipeline(object):
            def open_spider(self,spider):
                print("open_spider")
                self.start=datetime.datetime.now()
                self.con=sqlite3.connect("flights.db")
                self.cursor=self.con.cursor()
                try:
                    self.cursor.execute("drop table flights")
                except:
                    pass
                sql="""
                create table flights (
                    ID varchar(8) primary key,
                    direction varchar(32),
                    flight varchar(32),
                    company varchar(256),
                    flightType varchar(64),
                    airport varchar(256),
                    departureTime varchar(32),
                    arrivalTime varchar(32))
                """
                self.cursor.execute(sql)

            def close_spider(self,spider):
                print("close_spider")
                print("Total ",spider.count)
                self.con.commit()
                self.con.close()
                end=datetime.datetime.now()
```

```
                    print("Elapsed",(end-self.start).seconds,"seconds")

                def process_item(self, item, spider):
                    ID=item["ID"]
                    flight=item["flight"]
                    direction=item["direction"]
                    company=item["company"]
                    flightType=item["flightType"]
                    airport=item["airport"]
                    departureTime=item["departureTime"]
                    arrivalTime=item["arrivalTime"]
                    try:
                        sql="""
                        insert into flights
                        (ID,flight,direction,company,flightType,airport,departureTime,arrivalTime)
                        values (?,?,?,?,?,?,?,?)
                        """
                        self.cursor.execute(sql,[ID,flight,direction,company,flightType,airport,departureTime,arrivalTime])
                    except:
                        pass
                    return item
```

（3）编写 settings.py

程序如下：

```
                # Configure item pipelines
                # See https://docs.scrapy.org/en/latest/topics/item-pipeline.html
                ITEM_PIPELINES = {
                    'demo.pipelines.FlightPipeline': 300,
                }
```

（4）编写 MySpider.py

编写爬虫程序如下：

```
                import scrapy
                from demo.items import FlightItem

                class MySpider(scrapy.Spider):
                    name = "mySpider"

                    def start_requests(self):
                        self.count=0
```

```python
            directions=["arrivals","departures"]
            url='https://zh.flightaware.com/live/airport/ZGSZ/'
            for direction in directions:
                request=scrapy.Request(url=url+direction, callback=self.parse)
                request.meta["direction"]=direction
                yield request

    def parse(self, response):
        print(response.url)
        direction=response.meta["direction"]
        html=response.body.decode()
        selector=scrapy.Selector(text=html)
        trs= selector.xpath("//table[contains(@class,'prettyTable') and contains (@class, 'fullWidth')]//tr")
        for tr in trs[2:]:
            try:
                tds=tr.xpath(".//td")
                company=tds[0].xpath("./span/@title").extract_first()
                flight=tds[0].xpath("./span/a/text()").extract_first()
                flightType=tds[1].xpath("./span/@title").extract_first()
                airport=tds[2].xpath(".//span[@dir='ltr'][position()=1]/text()").extract_first()
                departureTime=tds[3].xpath("./text()").extract_first()
                arrivalTime=tds[4].xpath("./text()").extract_first()
                self.count=self.count+1
                item=FlightItem()
                item["ID"]="%06d"%self.count
                item["direction"]=direction
                item["flight"]=flight.strip() if flight else ""
                item["company"]=company.strip() if company else ""
                item["flightType"]=flightType.strip() if flightType else ""
                item["airport"]=airport.strip() if airport else ""
                item["departureTime"]=departureTime.strip() if departureTime else ""
                item["arrivalTime"]=arrivalTime.strip() if arrivalTime else ""
                yield item
            except Exception as err:
                print(err)

            #换页地址 url
            url=""
                links= selector.xpath("//table[contains(@class,'prettyTable') and contains
```

```
                (@class, 'fullWidth')]/following-sibling::span//a")
                for link in links:
                    s=link.xpath("./text()").extract_first()
                    if s=="后 20 条":
                        href=link.xpath("./@href").extract_first()
                        url=response.urljoin(href)
                        break
                if url:
                    request=scrapy.Request(url=url, callback=self.parse)
                    request.meta["direction"]=direction
                    yield request
```

程序的启动部分包括 direction="arrivals"的进港航班与 direction="departures"的出港航班，因此使用下面循环创建两个 scrapy.Request：

```
                for direction in directions:
                    request=scrapy.Request(url=url+direction, callback=self.parse)
                    request.meta["direction"]=direction
                    yield request
```

使用 request.meta 把 direction 参数传递给 parse()中的 response 对象。

4.6.7 执行爬虫程序

编写 run.py 如下：

```
from scrapy import cmdline
cmdline.execute("Scrapy crawl mySpider -s LOG_ENABLED=False".split())
```

执行 run.py 开始爬取数据，在 83 秒内爬取到了 3265 条记录。下面是部分输出结果，从结果看到 arrivals 与 departures 的数据是交替爬取存储，这也正是 Scrapy 分布式的体现。

```
                open_spider
                https://zh.flightaware.com/live/airport/ZGSZ/arrivals
                https://zh.flightaware.com/live/airport/ZGSZ/departures
                https://zh.flightaware.com/live/airport/ZGSZ/arrivals?;offset=20;order=actualarrivaltime;
sort=DESC
                https://zh.flightaware.com/live/airport/ZGSZ/departures?;offset=20;order=actualdeparture-
time;sort=DESC
                https://zh.flightaware.com/live/airport/ZGSZ/arrivals?;offset=40;order=actualarrivaltime;
sort=DESC
                    ……
                https://zh.flightaware.com/live/airport/ZGSZ/arrivals?;offset=1340;order=actualarrivaltime;
sort=DESC
```

https://zh.flightaware.com/live/airport/ZGSZ/departures?;offset=1840;order=actualdeparturetime;sort=DESC

https://zh.flightaware.com/live/airport/ZGSZ/departures?;offset=1860;order=actualdeparturetime;sort=DESC

https://zh.flightaware.com/live/airport/ZGSZ/departures?;offset=1880;order=actualdeparturetime;sort=DESC

close_spider

Total   3265

Elapsed 83 seconds

数据库表 flights 存储的航班数据，如图 4-6-4 所示。

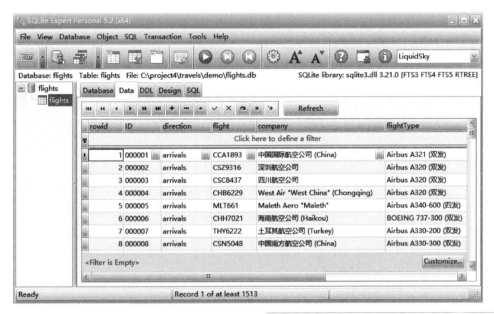

图 4-6-4 航班数据

练习四

① 说明 Scrapy 的入口地址的规则。

② 比较 Xpath 和 BeautifulSoup 解析 HTML 网页的特点与区别。

③ 在 Scrapy 中如何把爬取的数据写入数据库？items.py 与 pipelines.py 文件有什么用？

④ Python 中 yield 语句怎样工作？为什么 Scrapy 爬到的数据使用 yield 返回而不是使用 return 返回？

⑤ Scrapy 是如何实现多个网页爬取数据的？如何理解分布式爬取过程？

⑥ 仿照本书项目 1 的服务器程序，根据请求页面的不同展示 3 组学生的表格，这些学生分别存储在 students1.txt、students2.txt、students3.txt 文件中，每个页面结构相同（学号 No、姓名 Name、性别 Gender、年龄 Age），使用参数 page=N 控制展示 studentsN.txt 的学生表格，程序如下：

```python
import flask
app=flask.Flask(__name__)
@app.route("/")
def show():
    page = flask.request.args.get("page") if "page" in flask.request.args else "1"
    maxpage=3
    page=int(page)
    st="<h3>学生信息表</h3>"
    st=st+"<table border='1' width='300'>"
    fobj=open("students"+str(page)+".txt","rt",encoding="utf-8")
    while True:
        #读取一行，去除行尾"\n"换行符号
        s=fobj.readline().strip("\n")
        #如果读到文件尾部就退出
        if s=="":
            break
        #按逗号拆分开
        s=s.split(",")
        st=st+"<tr>"
        #把各个数据组织在<td>...</td>的单元中
        for i in range(len(s)):
            st=st+"<td>"+s[i]+"</td>"
        #完成一行
        st=st+"</tr>"
    fobj.close()
    st=st+"</table>"
    st=st+"<div>"
    if page>1:
        st = st + "<a href='/?page=" +str(page-1) + "'>【前一页】</a>"
    if page<maxpage:
        st=st+"<a href='/?page="+str(page+1)+"'>【下一页】</a>"
    st=st+"</div>"
    st=st+"<input type='hidden' name='page' value='"+str(page)+"'>"
    st=st+"<div>Page: "+str(page)+"/"+str(maxpage)+"</div>"
    return st

if __name__=="__main__":
    app.run()
```

服务器程序，如图 4-7-1 所示。

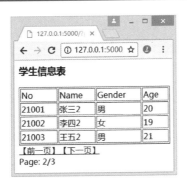

图 4-7-1
学生信息

使用 Scrapy 框架设计一个爬虫程序，爬取所有学生的信息并存储到数据库中。

项目 5　爬取商城网站数据

在实际应用中，很多网站的数据不是静态地嵌入 HTML 网页中的，而是通过 JavaScript 动态嵌入 HTML 网页中的。对于这类网站，使用普通的爬虫程序不能爬取网站的数据，而必须使用一种能执行网页中 JavaScript 程序的工具，才能爬取到网站的数据。Selenium 就是这样一种能模拟浏览器执行 JavaScript 的工具。在这个项目中使用 Selenium 编写爬虫程序去爬取某商城的商品数据。

5.1 使用 Selenium 编写爬虫程序

5.1.1 JavaScript 控制网页

网页上的信息不一定都是静态的 HTML 数据,实际上很多信息是通过 JavaScript 处理后得到的。那么,怎么样去爬取这些数据呢?本节先来设计一个由 JavaScript 动态控制的网页,看看怎么样爬取它的数据。

(1)创建网页模板

创建一个项目,在下面的 templates 文件夹中设计一个网页模板文件 notebook.html,它包含 3 个<div>:第一个 ID 是 hMsg,该信息是确定静态信息;第二个 ID 是 jMsg,该信息是在网页加载时由 JavaScript 的程序赋值"JavaScript 信息";第三个 ID 是 sMsg,该信息是网页在加载时通过 Ajax 的方法向服务器提出 GET 请求获取,服务器返回字符串"Ajax 服务器信息"。网页模板文件 notebook.html 如下:

```html
<script>
    function init()
    {
        http=new XMLHttpRequest();
        http.open("get","/ajax",false);
        http.send(null);
        msg=http.responseText;
        document.getElementById("sMsg").innerHTML=msg;
        document.getElementById("jMsg").innerHTML="JavaScript 信息";
    }
</script>
<body onload="init()">
Testing<br>
<div id="hMsg">静态信息</div>
<div id="jMsg"></div>
<div id="sMsg"></div>
</body>
```

(2)创建服务器程序

创建服务器程序 server.py,以显示出 notebook.html 文件的内容。其中,index()函数读取该文件并发送;show()函数是在接收地址/ajax 请求后发送"Ajax 服务器信息"。服务器程序 server.py 如下:

```python
import flask
app=flask.Flask(__name__)
@app.route("/")
def index():
    return flask.render_template("notebook.html")
```

```
@app.route("/ajax")
def ajax():
    return "Ajax 服务器信息"
app.run()
```

（3）浏览器浏览

启动服务器 server.py 程序，浏览 Web 地址 http://127.0.0.1:5000，如图 5-1-1 所示。

图 5-1-1
测试网站

5.1.2 普通爬虫程序问题

（1）普通客户端程序

编写爬虫程序 spider.py，通过 urllib.request 直接访问 http://127.0.0.1:5000，程序如下：

```
import urllib.request
resp=urllib.request.urlopen("http://127.0.0.1:5000");
data=resp.read()
data=data.decode()
print(data)
```

程序运行结果：

```
<script>
    function init()
    {
        http=new XMLHttpRequest();
        http.open("get","/ajax",false);
        http.send(null);
        msg=http.responseText;
        document.getElementById("sMsg").innerHTML=msg;
        document.getElementById("jMsg").innerHTML="JavaScript 信息";
    }
</script>
<body onload="init()">
Testing<br>
<div id="hMsg">静态信息</div>
<div id="jMsg"></div>
```

```
            <div id="sMsg"></div>
        </body>
```

该结果就是 notebook.html 文件。注意该结果中没有 id="jMsg" 与 id="sMsg" 的 <div> 信息，这些信息在程序运行后产生。

（2）编写普通爬虫程序

编写爬虫程序 spider.py，通过 urllib.request 直接访问 http://127.0.0.1:5000 获取 HTML 代码，使用 BeautifulSoup 解析得到数据，程序如下：

```
from bs4 import BeautifulSoup
import urllib.request
resp=urllib.request.urlopen("http://127.0.0.1:5000")
html=resp.read()
html=html.decode()
soup=BeautifulSoup(html,"lxml")
hMsg=soup.find("span",attrs={"id":"hMsg"}).text
print(hMsg)
jMsg=soup.find("span",attrs={"id":"jMsg"}).text
print(jMsg)
sMsg=soup.find("span",attrs={"id":"sMsg"}).text
print(sMsg)
```

程序运行结果：

```
静态信息
```

显然，通过该方法获取网页 HTML 文档然后进行数据爬取，只能爬取 hMsg 的信息 "静态信息"，而爬取不到 jMsg 与 sMsg 的信息。因为 jMsg 与 sMsg 的信息不是静态地嵌入在网页中的，而是通过 JavaScript 与 Ajax 动态产生的，但通过 urllib.request.urlopen 得到的网页中没有这样的动态信息。

5.1.3 安装 Selenium 框架

要获取 jMsg 与 sMsg 的这些信息，就必须在获取网页后客户端能够按要求执行对应的 JavaScript 程序。显然，一般的客户端程序没有这个能力去执行 JavaScript 程序，而必须寻找一个类似浏览器访问流程的库来完成该工作，Selenium 框架就是这样的库。

Selenium 能与通用的浏览器（如 Chrome、Firefox 等）配合工作，在本项目中把 Selenium 与 Chrome 配合，因此要安装 Selenium 与 Chrome 的驱动程序。

（1）安装 Selenium

```
pip install selenium
```

执行该命令即可完成安装。

（2）安装 Chrome 驱动程序

安装 Chrome 时要先查看 Chrome 的版本号，在 Chrome 浏览器的地址栏输入 about:version 就可以看到版本号，如图 5-1-2 所示。

图 5-1-2 浏览器版本号

然后在官方网站下载对应版本号的驱动程序 chromedrive.exe，下载完成后，将该驱动程序文件复制到 Python 的 scripts 目录下。

5.1.4 编写 Selenium 爬虫程序

（1）使用 Selenium 查看网页代码

按下列步骤编写客户端程序：

① 程序先从 Selenium 引入 webdriver，并引入 Chrome 程序的选择项目 Options：

```
from selenium import webdriver
from selenium.webdriver.chrome.options import Options
```

② 设置启动 Chrome 时不可见：

```
options.add_argument('--headless')
options.add_argument('--disable-gpu')
```

③ 创建 Chrome 浏览器：

```
driver= webdriver.Chrome(options=options)    #这样创建的 Chrome 浏览器是不可见的
```

④ 使用 driver.get(url) 方法访问网页：

```
driver.get("http://127.0.0.1:5000")
```

⑤ 通过 driver.page_source 获取网页 HTML 代码：

```
html=driver.page_source
print(html)
```

⑥ 使用 driver.close() 关闭浏览器：

```
driver.close()
```

根据这样的规则，编写爬虫程序 spider.py 如下：

```
from selenium import webdriver
from selenium.webdriver.chrome.options import Options
options = Options()
options.add_argument('--headless')
options.add_argument('--disable-gpu')
driver= webdriver.Chrome(options=options)
```

```python
driver.get("http://127.0.0.1:5000")
html=driver.page_source
print(html)
driver.close()
```

程序运行结果：

```html
<html><head><script>
    function init()
    {
        http=new XMLHttpRequest();
        http.open("get","/ajax",false);
        http.send(null);
        msg=http.responseText;
        document.getElementById("sMsg").innerHTML=msg;
        document.getElementById("jMsg").innerHTML="javascript 信息";
    }
</script>
</head><body onload="init()">
Testing<br>
<div id="hMsg">静态信息</div>
<div id="jMsg">JavaScript 信息</div>
<div id="sMsg">Ajax 服务器信息</div>
```

可见，得到的 HTML 文档是执行完 JavaScript 程序后的文档，其中包含了 jMsg 与 sMsg 的信息。

（2）编写 Selenium 爬虫程序

Selenium 模拟浏览器访问网站的方法来获取网页文档，然后从中爬取需要的数据，这样的爬虫程序功能强大了许多，编写爬虫程序 spider.py 如下：

```python
from selenium import webdriver
from selenium.webdriver.chrome.options import Options
from bs4 import BeautifulSoup
options = Options()
options.add_argument('--headless')
options.add_argument('--disable-gpu')
driver= webdriver.Chrome(options=options)
driver.get("http://127.0.0.1:5000")
html=driver.page_source
soup=BeautifulSoup(html,"lxml")
hMsg=soup.find("div",attrs={"id":"hMsg"}).text
print(hMsg)
jMsg=soup.find("div",attrs={"id":"jMsg"}).text
```

```
print(jMsg)
sMsg=soup.find("div",attrs={"id":"sMsg"}).text
print(sMsg)
driver.close()
```

程序运行结果：

```
静态信息
JavaScript 信息
Ajax 服务器信息
```

可见，使用 Selenium 的结构主要是模拟浏览器访问网页，并执行网页中的 JavaScript 程序，使得网页的数据充分下载，之后再用爬虫程序去爬取数据就稳妥了。

5.2 使用 Selenium 查找 HTML 元素

实际上 Selenium 支持 Xpath、CSS 等多种解析数据的方法，不需要再使用 BeautifulSoup 等工具去解析数据。

5.2.1 创建产品网站

（1）创建网页模板

在 templates 文件夹中编写一个 notebook.html 网页模板文件，内容如下：

```html
<div class="info">
    <div class="title"><h3 id="title" style="display:inline-block">笔记本电脑</h3></div>
    <div class="mark">
        <span class="pl">品牌</span>:<span name="mark">联想</span>
    </div>
    <div class="date">
        <span class="pl">生产日期</span>:<span name="date">2018-12-11</span>
    </div>
    <div class="price">
        <span class="pl">价格</span>:<span name="price">¥6200.00</span>
    </div>
    <div><a href='/brief'>产品简介</a></div>
    <div><a href='/details'>产品详情</a></div>
</div>
```

（2）创建网站服务器程序

创建一个服务器程序 server.py 用来展示 notebook.html 网页，程序如下：

```python
import flask
app=flask.Flask(__name__)
@app.route("/")
```

```
        def show():
            return flask.render_template("notebook.html")
        app.run()
```

执行服务器程序，使用浏览器浏览 http:/127.0.0.1:5000，如图 5-2-1 所示。

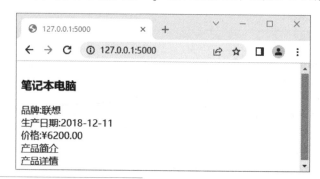

图 5-2-1
产品网站

（3）创建查找程序

Selenium 搭配有很多种查找 HTML 元素的方法，通过这些方法就可以爬取到所要的数据，为了方便说明，编写下列程序：

```
from selenium import webdriver
from selenium.webdriver.chrome.options import Options
options = Options()
options.add_argument('--headless')
options.add_argument('--disable-gpu')
driver= webdriver.Chrome(options=options)
driver.get("http://127.0.0.1:5000")
#查找元素与数据
driver.close()
```

Selenium 查找元素一般使用 find_element()方法，函数使用方法如下：

```
find_element(by,value)
```

其中 by 是通过什么方法查找，by 的值通过下面的语句引入：

```
from selenium.webdriver.common.by import By
```

例如，使用 Xpath 方法查找，则设置 by=By.XPATH；使用 ID 号查找，则设置 by=By.ID。value 是要查找元素的匹配值。不同的查找方法，该值不同。

5.2.2 使用 Xpath 查找元素

使用 Xpath 查找主要有两个函数：

① 函数 find_element(by=By.XPATH,value)：查找匹配的第一个元素，如果找到就返回一个 WebElement 类的对象，如果找不到就抛出异常。

② 函数 find_elements(by=By.XPATH,value)：查找匹配的所有元素组成的列表，每个

元素都是一个 WebElement 对象，如果找不到就返回空列表。

WebElement 是一个 Selenium 中定义的类，任何一个 WebElement 对象都可以再次调用 find_element()与 find_elements()函数。

例 5-2-1： 查找网页中<div class="info">元素

```
elem=driver.find_element(by=By.XPATH,value="//div[@class='info'] ")
print(type(elem))
```

程序运行结果：

```
<class 'selenium.webdriver.remote.webelement.WebElement'>
```

例 5-2-2： 查找网页中元素<h4>

```
try:
    elem=driver.find_element(by=By.XPATH,value="//div[@class='info']//h4")
    print(type(elem))
except Exception as err:
    print(err)
```

结果发现没有找到<h4>，抛出下列异常错误信息：

```
Message: no such element: Unable to locate element: {"method":"xpath","selector":"//div[@class='info']//h4"}
```

5.2.3 查找元素的文本与属性

通过 WebElement 对象可以查找到它的文本与属性。

① 任何一个 WebElement 对象都可以通过 text 属性获取它的文本，元素的文本值是它与它的所有子孙节点的文字的组合，如果没有就返回空字符串。

② 任何一个 WebElement 对象都可以通过 get_attribute(attrName)获取名称为 attrName 的属性值，如果元素没有 attrName 属性就返回 None。

例 5-2-3： 查找生产日期

```
elem=driver.find_element(by=By.XPATH,value="//div[@class='date']//span[@name='date']")
print(elem.text)
```

程序运行结果：

```
2018-12-11
```

例 5-2-4： 查找网页中<div class='mark'>中所有<span>元素的文本

```
elem=driver.find_element(by=By.XPATH,value="//div[@class='mark']")
elems=elem.find_elements (by=By.XPATH,value=".//span")
for elem in elems:
    print(elem.text)
```

或：

```
elems=driver.find_elements(by=By.XPATH,value="//div[@class='mark']//span")
for elem in elems:
    print(elem.text)
```

程序运行结果：

```
品牌
联想
```

例 5-2-5： 查找网页中笔记本的品牌

```
print(driver.find_element (by=By.XPATH,value="//div[@class='info']//span [@name='mark']").text)
```

程序运行结果：

```
联想
```

例 5-2-6： 查找网页中<h3>的样式

```
elem=driver.find_element(by=By.XPATH,value="//h3")
print(elem.get_attribute("style"))
```

程序运行结果：

```
display: inline-block;
```

例 5-2-7： 查找网页中全部具有 name 属性的<span>元素的文本

```
elems=driver.find_elements(by=By.XPATH,value="//div//span[@name]")
for elem in elems:
    print(elem.text)
```

程序运行结果：

```
联想
2018-12-11
¥6200.00
```

例 5-2-8： 查找网页中<div class='mark'>的 HTML 文本

```
elem=driver.find_element(by=By.XPATH,value="//div[@class='mark']")
print("innerHTML")
print(elem.get_attribute("innerHTML").strip())
print("outerHTML")
print(elem.get_attribute("outerHTML").strip())
```

程序运行结果：

```
innerHTML
```

```
<span class="pl">品牌</span>:<span name="mark">联想</span>
outerHTML
<div class="mark">
    <span class="pl">品牌</span>:<span name="mark">联想</span>
</div>
```

例 5-2-9： 查找所有超链接

```
elems=driver.find_elements(by=By.XPATH,value="//a")
for elem in elems:
    print(elem.get_attribute("href"))
```

程序运行结果：

```
http://127.0.0.1:5000/brief
http://127.0.0.1:5000/details
```

5.2.4 使用 id 查找元素

HTML 中很多元素都有一个唯一的 id 值，Selenium 可以通过 id 值查找到元素。

函数 driver.find_element(by=By.ID,value)查找匹配编号的第一个元素，如果查找到就返回一个 WebElement 对象，如果没有找到就抛出异常。

例 5-2-10： 查找网页中 id="title"的元素文本

```
elem=driver.find_element(by=By.ID,value="title")
```

程序运行结果：

```
笔记本电脑
```

例 5-2-11： 查找网页中 id="name"的元素

```
elem=driver.find_element(by=By.ID,value="name")
print(elem.text)
```

程序运行结果，抛出异常：

```
Message: no such element: Unable to locate element: {"method":"css selector", "selector":"[id="name"]"}
```

5.2.5 使用 name 查找元素

HTML 中很多元素都有一个 name 属性值，Selenium 可以通过 name 属性值查找到元素。

① 函数 find_elemen(by=By.NAME,value)：查找匹配的第一个元素，如果找到就返回一个 WebElement 类型的对象，如果找不到就抛出异常。

② 函数 find_elements(by=By.NAME,value)：查找匹配的所有元素组成的列表，每个元素都是一个 WebElement 对象，如果找不到就返回空列表。

例 5-2-12： 查找网页中产品品牌

```
print(driver.find_element(by=By.NAME,cvalue="mark").text)
```

程序运行结果：

联想

例 5-2-13： 查找网页 name="xxx"的元素

```
try:
    driver.find_element_by_name("xxx")
except Exception as err:
    print(err)
```

程序运行结果，抛出异常：

Message: no such element: Unable to locate element: {"method":"name","selector":"xxx"}

5.2.6 使用 CSS 查找元素

Selenium 也支持 CSS 语法查找元素。

① 函数 find_element(by=By.CSS_SELECTOR,css)：查找 CSS 匹配的第一个元素，如果找到就返回一个 WebElement 类型的对象，如果找不到就抛出异常。

② 函数 find_elements_by_css_selector(css)：查找 CSS 匹配的所有元素组成的列表，每个元素都是一个 WebElement 对象，如果找不到就返回空列表。

例 5-2-14： 查找网页中产品价格

```
elem=driver.find_element(by=By.CSS_SELECTOR,value="div[class='price'], span[name='price']")
```

程序运行结果：

¥6200.00

例 5-2-15： 查找网页中产品图像地址

```
print(driver.find_element_by_css_selector("div[class='mark']>div").get_attribute("src"))
```

程序运行结果：

http://127.0.0.1:5000/images/000001.jpg

例 5-2-16： 查找网页中\<div class='price'>下面的所有元素

```
elems=driver.find_elements(by=By.CSS_SELECTOR,value="div[class='price'] *")
for elem in elems:
    print(elem.text)
```

程序运行结果：

价格
¥6200.00

例 5-2-17： 查找网页中笔记本电脑

```
elem=driver.find_element(by=By.CSS_SELECTOR,value="[id='title']")
```

或：

```
elem=driver.find_element(by=By.CSS_SELECTOR,value="#title")
```

程序运行结果：

```
笔记本电脑
```

例 5-2-18： 查找网页中所有具有 name 属性的<span>元素的文本

```
elems=driver.find_elements(by=By.CSS_SELECTOR,value="span[name]")
for elem in elems:
    print(elem.text)
```

程序运行结果：

```
联想
2018-12-11
¥6200.00
```

5.2.7 使用标签查找元素

Selenium 还可以通过 HTML 元素的 tagName 查找。

函数 find_elements(by=By.TAG_NAME,tagName)：查找 tagName 匹配的所有元素，如果找到就返回一个 WebElement 列表，如果找不到列表为空。

例 5-2-19： 查找<div class='price'>元素下面的所有<span>元素

```
elem=driver.find_element(by=By.XPATH,value="//div[@class='price']")
elems=elem.find_elements(by=By.TAG_NAME,value="span")
for elem in elems:
    print(elem.text)
```

程序运行结果：

```
价格
¥6200.00
```

5.2.8 查找超链接

Selenium 可以通过超级链接的文本来查找到该超链接。

① 函数 find_element(by= By.LINK_TEXT,value)查找第一个文本值为 text 的超链接元素<a>，如果找到就返回该元素的 WebElement 对象，如果找不到就抛出异常。

② 函数 find_element(by= By.PARTIAL_LINK_TEXT,value)查找第一个文本值包含 value 的超链接元素<a>，如果找到就返回该元素的 WebElement 对象，如果找不到就抛出异常。

③ 函数 find_elements(by= By.LINK_TEXT,value)查找所有文本值为 value 的超链接元素<a>，如果找到就返 WebElement 列表，如果找不到列表为空。

④ 函数 find_elements(by= By.PARTIAL_LINK_TEXT,value)查找所有文本值包含 value 的超链接元素<a>，找到就返 WebElement 列表，如果找不到列表为空。

例 5-2-20: 查找网页中有"详情"字样的<a>元素 href：

```
elem=driver.find_element(by=By.LINK_TEXT,value="产品详情")
print(elem.get_attribute("href"))
elem=driver.find_element(by=By.PARTIAL_LINK_TEXT,value="详情")
print(elem.get_attribute("href"))
```

几种方法都能找到，程序运行结果：

```
http://127.0.0.1:5000/details
http://127.0.0.1:5000/details
```

但是 driver.find_element(by=By.LINK_TEXT",value="详情")是找不到的，因为这个函数要求文本完全匹配。

5.2.9 使用 class 查找元素

（1）查找单一类名的元素

Selenium 可以使用元素的 class 值查找元素。

① 函数 find_element(by=By.CLASS_NAME,value)查找第一个 class=value 的元素，如果找到就返回该元素的 WebElement 对象，如果找不到就抛出异常。

② 函数 find_elements(by=By.CLASS_NAME,value)查找所有 class=value 元素，如果找到就返 WebElement 列表，如果找不到列表为空。

例 5-2-21: 查找网页 class="pl"的所有元素

```
elems=driver.find_elements(by=By.CLASS_NAME,value="pl")
for elem in elems:
    print(elem.text)
```

程序运行结果：

```
品牌
生产日期
价格
```

也可以通过以下方式查找：

```
elems=driver.find_elements(by=By.XPATH,value="//*[@class='pl']")
elems=driver.find_elements(by=By.CSS_SELECTOR,value="*[class='pl']")
```

（2）查找复合类名的元素

 注意：

在网页中有些元素有复合的类名称。例如，修改网页中的生产日期部分为：

```
<div class="date">
<span class="pl date">生产日期</span>:<span name="date">2016-12-01</span>
</div>
```

那么这个<span class="pl date">就有复合的 class 名称，一个是 pl，另一个是 date，两者的名称之间有一个空格。

这种类型的元素不能使用 find_element(by=By.CLASS_NAME,value)方法查找，但是可以使用 Xpath 或者 CSS 语法的函数查找。例如，在已知 pl 与 date 的空格数的情况下可以直接使用以下方法之一：

```
driver.find_element(by=By.XPATH,"//span[@class='pl date']")
driver.find_element(by=By.CSS_SELECTOR,value="span[class='pl date']")
```

通常 pl 与 date 之间的空格数会不确定，那么更加通用的匹配方法如下：

```
elem=driver.find_element(by=By.XPATH,value="//span[contains(@class,'pl') and contains(@class,'date')]")
elem=driver.find_element(by=By.CSS_SELECTOR,value="span[class*='pl'][class*='date']")
```

5.3 使用 Selenium 实现用户登录

很多网站都需要用户登录后才能访问到其他网页，这个过程可以使用 Selenium 自动完成。Selenium 查找的 HTML 元素是一个 WebElement 对象，这个对象不但可以获取元素的属性值，而且还能执行一些键盘输入与鼠标单击的动作。

5.3.1 创建用户登录网站

（1）创建网页模板

在 static 文件夹下面放一张图像 demo.jpg，在 templates 中创建一个 login.html 的用户登录网页模板文件，在启动时显示登录界面，用户输入用户名 xxx 及密码 123 后可完成登录，然后转到/show 的页面显示 demo.jpg 图像，login.html 内容如下：

```
<body>
<form id="frm" action="/" method="post">
<div>用户<input type="text" name="user" id="user"></div>
<div>密码<input type="password" name="pwd" id="pwd"></div>
<div><input type="submit" name="login" id="login" vaule='登录'></div>
</form>
</body>
```

（2）创建服务器程序

服务器程序 server.py 首先提交一个 login.html 的网页，用户输入名称与密码后提交，服务器获取用户名称与密码后判断是否正确，若不正确就继续显示登录页面，若正确就转/show 页面显示产品记录，服务器程序如下：

```
import flask
```

```python
app=flask.Flask(__name__)

@app.route("/",methods=["GET","POST"])
def login():
    flask.session["login"]=""
    if flask.request.method=="POST":
        user=flask.request.values.get("user","")
        pwd=flask.request.values.get("pwd","")
        if user=="xxx" and pwd=="123":
            flask.session["login"]="OK"
            return flask.redirect("/show")
    return flask.render_template("login.html")

@app.route("/show",methods=["GET","POST"])
def show():
    if flask.session.get("login","")=="OK":
        return "<img src='/static/demo.jpg'>"
    return flask.redirect("/")

app.secret_key="123"
app.debug=True
app.run()
```

（3）浏览器浏览

启动服务器程序后浏览 Web 地址 http://127.0.0.1:5000，如图 5-3-1 所示。用户登录成功后就可以看到图像。

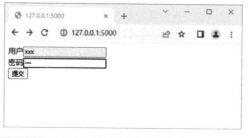
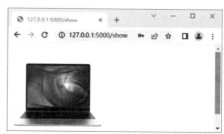

图 5-3-1 登录成功

显然，如果要爬取 Web 网站上的这张图像，就必须模拟用户登录，成功后访问网址为 http://127.0.0.1:5000/show 的页面。

5.3.2 使用元素动作

Selenium 查找 HTML 时返回一个 WebElement 对象，该对象功能十分强大，不但可以使用它获取元素的属性值，还可以模拟动作，例如键盘输入动作与鼠标单击动作。

（1）键盘输入动作

用户在元素<input type="text">文本输入框中输入文字，WebElement 对象可以模拟用户的键盘输入动作，主要动作有：

① 函数 clear() 模拟清除元素中的所有文字。

② 函数 send_keys(string) 模拟键盘在元素中输入字符串 string。

send_keys() 函数不但可以模拟输入文字，而且还可以模拟按回车、空格等键，Selenium 提供了一个 Keys 类，包含了很多常用的特殊按键，例如 Keys.BACKSPACE（Backspace 键）和 Keys.ENTER（回车键）。

编写 spider.py 程序如下：

```python
from selenium import webdriver
from selenium.webdriver.common.by import By
import time
try:
    driver = webdriver.Chrome()
    driver.get("http://127.0.0.1:5000")
    user = driver.find_element(by=By.NAME,value="user")
    pwd = driver.find_element(by=By.NAME,value="pwd")
    time.sleep(0.5)
    user.send_keys("xxx")
    time.sleep(0.5)
    pwd.send_keys("123")
except Exception as err:
    print(err)
input("Strike any key to finish...")
driver.close()
```

使用 driver = webdriver.Chrome() 创建浏览器，因此在浏览器中的模拟动作是可见的，运行服务器程序后执行该程序，可以清楚地看见程序在"用户"文本输入框与"密码"文本输入框的自动输入动作，如图 5-3-2 所示。

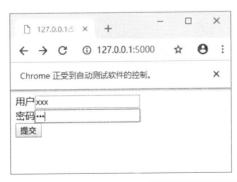

图 5-3-2
实现键盘自动输入

（2）鼠标单击动作

元素 \<input type="submit"> 需要鼠标单击动作，提交按钮单击后就提交表单。WebElement 使用 click() 函数模拟鼠标单击，例如：

```
driver.find_element (by=By.XPATH,"//input[@type='submit']").click()
```

5.3.3 编写爬虫程序

使用 Selenium 模拟用户使用浏览器进行浏览与登录，爬取图像，程序过程如下：

① 创建一个浏览器对象 driver，使用 driver 对象模拟浏览器。

② 访问 http://127.0.0.1:5000 网站，获取<input type="text" name="user">与<input type="password" name="pwd">元素对象，模拟用户在"用户"输入框输入 xxx，在"密码"输入框输入 123。

③ 获取<input type="submit" name="login">按钮对象，执行 click()模拟单击动作提交表单。

④ 服务器接收提交的 user 与 pwd 数据，判断是否登录成功，如果登录成功就转 http://127.0.0.1:5000/show 页面显示产品记录。

⑤ 爬取并下载图像：

编写爬虫程序 spider.py 如下。

```python
from selenium import webdriver
from selenium.webdriver.common.by import By
from selenium.webdriver.chrome.options import Options
import time
import urllib.request

def download(src):
    print(src)
    try:
        response=urllib.request.urlopen(src)
        data=response.read()
        p=src.rfind("/")
        fn=src[p+1:]
        fobj=open(fn,"wb")
        fobj.write(data)
        fobj.close()
        print("downloaded",src)
    except Exception as err:
        print(err)

def login():
    user = driver.find_element(by=By.NAME,value="user")
    pwd = driver.find_element(by=By.NAME,value="pwd")
    login = driver.find_element(by=By.NAME,value="login")
    user.send_keys("xxx")
    pwd.send_keys("123")
    login.click()
```

```
            time.sleep(0.5)
        try:
            url="http://127.0.0.1:5000"
            options = Options()
            options.add_argument('--headless')
            options.add_argument('--disable-gpu')
            driver = webdriver.Chrome(options=options)
            driver.get(url)
            login()
            elem=driver.find_element(by=By.XPATH,value="//img")
            src=elem.get_attribute("src")
            src=urllib.request.urljoin(url,src)
            download(src)
        except Exception as err:
            print(err)
        driver.close()
```

程序先访问 http://127.0.0.1:5000，user.send_keys("xxx")与 pwd.send_keys("123")是向 user 与 pwd 元素发送文本字符串 xxx 与 123。

login.click()是模拟单击"提交"按钮的动作，执行该命令即完成提交，提交后网页转到 http://127.0.0.1:5000/show，使用 time.sleep(0.5)停顿一会儿，等待网页加载完成后开始下载图像，程序运行结果：

```
http://127.0.0.1:5000/static/demo.jpg
downloaded http://127.0.0.1:5000/static/demo.jpg
```

5.3.4 执行 JavaScript 程序

（1）无参数的 JavaScript 程序

实际上这个登录过程也可以用 JavaScript 程序实现，使用 document 中的 getElementById()方法找到 user 与 pwd 的对象，分别设置值为 xxx 与 123，再找到 login 对象并调用 click()函数就完成登录。

Selenium 使用 execute_script(js)方法执行 JavaScript 程序，其中 js 是要执行的 JavaScript 程序代码，因此 Selenium 使用 JavaScript 编写功能完全一样的 login()函数如下：

```
        def login():
            js="""
            document.getElementById('user').value='xxx';
            document.getElementById('pwd').value='123';
            document.getElementById('login').click();
            """
            driver.execute_script(js)
```

```
                time.sleep(0.5)
```

（2）有参数的 JavaScript 程序

Selenium 在执行 execute_script(js)时还可以使用 Selenium 的 WebElement 对象参数代替 JavaScript 中的 DOM 对象参数，这个时候在 JS 中使用 arguments[n]形式参数，其中 n 表示参数的顺序。

例如使用 Selenium 找到 user 对象并设置它的值为 xxx 的语句如下：

```
                user=driver.find_element(by=By.ID,value="user")
                driver.execute_script("arguments[0].value='xxx'",user)
```

这里的 arguments[0]是参数，执行时被 WebElement 对象 user 对象代替，效果和如下语句一致：

```
                driver.execute_script("document.getElementById('user').value='xxx'")
```

同时为 user 与 pwd 设置值可以使用下面程序：

```
                user=driver.find_element(by=By.ID,value="user")
                pwd = driver.find_element(by=By.ID,value="pwd")
                driver.execute_script("arguments[0].value='xxx';",user)
                driver.execute_script("arguments[0].value='123';",pwd)
```

也可以合并在一起执行，下面的程序效果与上述程序效果一致。

```
                user=driver.find_element(by=By.ID,value="user")
                pwd = driver.find_element(by=By.ID,value="pwd")
                driver.execute_script("arguments[0].value='xxx';arguments[1].value='123';",user,pwd)
```

在执行 execute_script()时有两个参数 arguments[0]与 arguments[1]，执行时被实际的 user 与 pwd 代替。

根据这些规则，login()函数可以重新编写成如下形式：

```
                def login():
                    user = driver.find_element(by=By.NAME,value="user")
                    pwd = driver.find_element(by=By.NAME,value="pwd")
                    login = driver.find_element(by=By.NAME,value="login")
                    js="arguments[0].value='xxx';arguments[1].value='123';arguments[2].click();"
                    driver.execute_script(js,user,pwd,login)
                    time.sleep(0.5)
```

5.4 使用 Selenium 爬取 Ajax 网页数据

网页中大量使用了 Ajax 技术，通过在客户端的 JavaScript 向服务器发出请求，服务器返回数据给客户端，客户端展现数据，这样做可以减少网页的闪动，提升用户体验。

5.4.1 创建 Ajax 网站

（1）创建服务器程序

服务器程序 server.py 首先提交一个 notebook.html 的网页，然后响应/items?mark=...的请求，根据 mark 的值确定品牌，返回该品牌下的产品记录，返回的记录采用 JSON 数据格式，服务器程序如下：

```python
import flask
import json
app=flask.Flask(__name__)

@app.route("/")
def index():
    return flask.render_template("notebook.html")


@app.route("/items")
def getItems():
    mark=flask.request.values.get("mark","")
    items=[]
    if mark=="华为":
        items.append({"model":"MateBook D 14","mark":"华为","price":5000})
        items.append({"model": "MateBook D 14 SE", "mark": "华为", "price": 4000})
    elif mark=="联想":
        items.append({"model":"ThinkPad E14","mark":"联想","price":6800})
        items.append({"model":"Air14Plus","mark":"联想","price":3900})
    elif mark=="戴尔":
        items.append({"model":"Dell 3400","mark":"戴尔","price":5800})
    s=json.dumps({"items":items})
    return s

app.debug=True
app.run()
```

（2）创建网页文件

在 templates 文件夹中创建网页文件 notebook.html，内容如下。

```html
<script>
    function init()
    {
        var marks=["华为","联想","戴尔"];
        var selm=document.getElementById("marks");
        for(var i=0;i<marks.length;i++)
```

```javascript
            {
                selm.options.add(new Option(marks[i],marks[i]));
            }
            selm.selectedIndex=0;
            display();
        }

        function display()
        {
            try
            {
                var http=new XMLHttpRequest();
                var selm=document.getElementById("marks");
                var m=selm.options[selm.selectedIndex].text;
                http.open("get","/items?mark="+m,false);
                http.send(null);
                msg=http.responseText;
                obj=eval("("+msg+")");
                s="<table width='200' border='1'><tr><td>型号</td><td>价格</td></tr>"
                for(var i=0;i<obj.items.length;i++)
                {
                    s=s+"<tr><td>"+obj.items[i].model+"</td><td>"+obj.items[i].price+"</td></tr>";
                }
                s=s+"</table>";
                document.getElementById("items").innerHTML=s;
            }
            catch(e) { alert(e); }
        }
    </script>
    <body onload="init()">
    <div>选择品牌<select id="marks" onchange="display()"></select></div>
    <div id="items"></div>
    </body>
```

程序说明：

① 网页框架：该网页的主体框架很简单，只有两个<div>元素，一个包含了<select>下拉菜单，另外一个是显示信息使用的，在没有执行 JavaScript 之前内容都是空的。

② init()函数：在网页被加载时执行 init()函数，为<select>初始设置了三个笔记本电脑品牌，即华为、联想、戴尔。

③ display()函数：当用户选择其中一个品牌时就触发<select>的 onchange 事件，从而执行 display()函数，该函数通过 Ajax 技术把选择的笔记本品牌通过 http.open("get","/items?mark="+m,false)语句发送给服务器，服务器收到该品牌后返回这个品牌的信息 http.responseText 给这个网页。返回的数据是 JSON 格式的字符串，经 eval()转换为 JavaScript 对象 items 后生成一张表格的 HTML 代码放到<div id="items">...</div>中显示。

（3）浏览器浏览

启动服务器程序浏览 Web 地址 http://127.0.0.1:5000，如图 5-4-1 所示。用户选择另外一个品牌后就触发 notebook.html 中<select>的 onchange 事件，再次使用 Ajax 获取该品牌的记录进行显示。

图 5-4-1
笔记本电脑网站

5.4.2 理解 Selenium 爬虫程序

（1）普通爬虫程序

编写下列爬虫程序 spider.py 获取网页的 HTML 代码：

```
import urllib.request
resp=urllib.request.urlopen("http://127.0.0.1:5000")
html=resp.read().decode()
print(html)
```

执行该程序可以看到 HTML 实际上就是原始的 notebook.html，如果要爬取这个 Web 网站的所有记录，使用简单的爬虫程序方法是行不通的，因为该网页 HTML 代码中根本就看不出任何的记录信息。

（2）Selenium 爬虫程序

Selenium 可以模拟浏览器，在浏览网页后会在内部构建一棵 HTML 元素 DOM 网页树，编写 spider.py 程序如下：

```
from selenium import webdriver
from selenium.webdriver.chrome.options import Options
try:
    url="http://127.0.0.1:5000"
    options = Options()
    options.add_argument('--headless')
    options.add_argument('--disable-gpu')
    driver = webdriver.Chrome(options=options)
```

```
            driver.get(url)
            print(driver.page_source)
        except Exception as err:
            print(err)
        driver.close()
```

程序运行结果：

```
<html><head><script>
    function init()
    {
        var marks=["华为","联想","戴尔"];
        var selm=document.getElementById("marks");
        for(var i=0;i<marks.length;i++)
        {
            selm.options.add(new Option(marks[i],marks[i]));
        }
        selm.selectedIndex=0;
        display();
    }

    function display()
    {
        try
        {
            var http=new XMLHttpRequest();
            var selm=document.getElementById("marks");
            var m=selm.options[selm.selectedIndex].text;
            http.open("get","/items?mark="+m,false);
            http.send(null);
            msg=http.responseText;
            obj=eval("("+msg+")");
            s="<table width='200' border='1'><tr><td>型号</td><td>价格</td></tr>"
            for(var i=0;i<obj.items.length;i++)
            {
                s=s+"<tr><td>"+obj.items[i].model+"</td><td>"+obj.items[i].price+"</td></tr>";
            }
            s=s+"</table>";
            document.getElementById("items").innerHTML=s;
        }
```

```
            catch(e) { alert(e); }
        }
    </script>
    </head><body onload="init()">
    <div>选择品牌<select id="marks" onchange="display()"><option value="华为">华为
</option><option value="联想">联想</option><option value="戴尔">戴尔</option>
</select></div>
    <div id="items"><table width="200" border="1"><tbody><tr><td>型号</td><td>价格
</td></tr><tr><td>MateBook D 14</td><td>5000</td></tr><tr><td>MateBook D 14 SE</td>
<td>4000</td></tr></tbody></table></div>
    </body></html>
```

可见，这棵树的元素包括静态 HTML 的元素，也包括从 JavaScript 执行后产生的元素，Selenium 会按这棵树去查找各个元素。

5.4.3 编写爬虫程序

使用 Selenium 模拟浏览器浏览网页，然后再模拟用户选择<select>中各个品牌的操作实现换页，就可以实现逐页爬取所有的数据了，爬虫程序流程如下。

① 创建一个浏览器对象 driver，使用这个 driver 对象模拟浏览器。
② 访问 http://127.0.0.1:5000 网站，爬取第一个页面的数据。
③ 从第一个页面中获取<select>中所有的选择项目 options。
④ 遍历 options 中的每个 option，并模拟对 option 执行单击动作，触发 onchange 事件调用 display()函数，爬取每个页面的数据。

根据这个规则，编写爬虫程序 spider.py 如下：

```python
from selenium import webdriver
from selenium.webdriver.chrome.options import Options
from selenium.webdriver.common.by import By
import time

def spider():
    trs=driver.find_elements(by=By.TAG_NAME,value="tr")
    for tr in trs[1:]:
        tds=tr.find_elements(by=By.TAG_NAME,value="td")
        model=tds[0].text
        price=tds[1].text
        print("%-16s%-16s" %(model,price))

options = Options()
options.add_argument('--headless')
options.add_argument('--disable-gpu')
```

```
            driver = webdriver.Chrome(options=options)
            driver.maximize_window()
            driver.get("http://127.0.0.1:5000")
            select = driver.find_element(by=By.ID,value="marks")
            options = select.find_elements(by=By.TAG_NAME,value="option")
            for option in options:
                option.click()
                time.sleep(0.1)
                spider()
            driver.close()
```

程序说明：

创建浏览器对象：

```
            driver = webdriver.Chrome(options=options)
```

浏览网站的第一页：

```
            driver.get("http://127.0.0.1:5000")
```

spider()函数负责爬取当前页面的所有项目记录，其中：

```
            trs=driver.find_elements(by=By.TAG_NAME,value="tr")
```

获取所用的<tr>元素，在网页中有很多个<tr>元素，然后通过循环去获取每个<tr>元素，程序跳过第一个<tr>的表格头，从第二个<tr>开设是产品的记录。通过：

```
            tds=trs.find_elements(by=By.TAG_NAME,value="td")
```

获取<tr>下面的所有<td>元素，每个<tr>下面有两个<td>元素，第一个是型号，第二个是价格：

```
            model=tds[0].text
            price=tds[1].text
```

程序通过：

```
            select = driver.find_element(by=By.ID,value="marks")
```

获取网页中的<select>元素后，通过：

```
            options = select.find_elements(by=By.TAG_NAME,value="option")
```

获取该元素下所有的<option>元素。

循环遍历 options 中每个元素：

```
            for option in options:
                option.click()
                time.sleep(0.1)
                spider()
```

为每个元素调用 click()函数，每次 option.click()都是一个模拟用户单击该<option>的动作，会触发<select>的 onchange 事件，从而执行 display()函数，用 Ajax 从服务器获取该记录，再次调用 spider()就可以爬取。

注意每次执行 option.click()后网页中的<select>元素整体是没有改变的，因此可以使用循环的方法连续调用 option.click()。

5.4.4 执行爬虫程序

执行该程序，爬取到了所有的产品记录：

MateBook D 14	5000
MateBook D 14 SE	4000
ThinkPad E14	6800
Air14Plus	3900
Dell 3400	5800

5.5 爬取网站换页数据

上一节案例中<select>每次变化时都通过 Ajax 获取一组新的数据，注意这个网页是没有更新的，只是网页中的数据更新，使用 Selenium 模拟<select>中<option>的变化爬取了全部的数据。但是，如果<option>变化会重新加载一个新的网页，那么，怎样去爬取全部数据呢？

5.5.1 创建实验网站

（1）创建服务器程序

服务器程序 server.py 首先提交一个 notebook.html 的网页,然后响应/?mark=...的请求,根据 mark 的值确定品牌，返回该品牌下的产品记录，服务器程序如下：

```
import flask
app=flask.Flask(__name__)

@app.route("/")
def index():
    mark=flask.request.values.get("mark","华为")
    items=[]
    if mark=="华为":
        items.append({"model":"MateBook D 14","mark":"华为","price":5000})
        items.append({"model": "MateBook D 14 SE", "mark": "华为", "price": 4000})
    elif mark=="联想":
        items.append({"model":"ThinkPad E14","mark":"联想","price":6800})
        items.append({"model":"Air14Plus","mark":"联想","price":3900})
```

```
            elif mark=="戴尔":
                items.append({"model":"Dell 3400","mark":"戴尔","price":5800})
            return flask.render_template("notebook.html",items=items,mark=mark)

app.debug=True
app.run()
```

（2）创建网页文件

在 templates 文件夹中创建网页文件 notebook.html，它的内容如下：

```
<script>
    function display()
    {
        var selm=document.getElementById("marks");
        var m=selm.options[selm.selectedIndex].text;
        window.location.href="?mark="+m;
    }
</script>
<body>
<div>选择品牌<select id="marks" onchange="display()">
<option value="华为" {% if mark=="华为" %}selected{% endif %}>华为</option>
<option value="联想" {% if mark=="联想" %}selected{% endif %}>联想</option>
<option value="戴尔" {% if mark=="戴尔" %}selected{% endif %}>戴尔</option>
</select></div>
<table width="300" border="1">
<tr><td>型号</td><td>价格</td></tr>
{% for item in items %}
<tr><td>{{item.model}}</td><td>{{item.price}}</td></tr>
{% endfor %}
</table>
</body>
```

当用户选择其中一个品牌时就触发<select>的 onchange 事件，从而执行 display()函数，该函数执行：

```
window.location.href="?mark="+m;
```

通过 mark 的值获取新的 items 数据，通过新网页展示。

（3）浏览器浏览

启动服务器程序浏览 Web 地址 http://127.0.0.1:5000，如图 5-5-1 所示。用户选择另外一个品牌后就触发 notebook.html 中<select>的 onchange 事件，获取该品牌的记录进行显示。

图 5-5-1
实验网站

5.5.2 爬虫程序问题

在前面的 Ajax 项目中编写的爬虫程序 spider.py，如果使用这个程序去爬取数据会出现下面的报错：

> Message: stale element reference: element is not attached to the page document

报错意思是当前的页面中没有绑定给定的元素，其原因请仔细分析下面的程序：

```
select = driver.find_element(by=By.ID,value="marks")
options = select.find_elements(by=By.TAG_NAME,value="option")
for option in options:
    option.click()
    time.sleep(0.1)
    spider()
```

实际上该报错是 option.click() 语句导致的，因为在第一个页面中就确定了 options 列表，然后执行 option.click()，这个动作会导致 display() 函数执行语句：

```
window.location.href="?mark="+m;
```

这条语句会开启一个新的页面，重新构建 <select> 中的 <option> 元素，新的 <option> 元素不再是原来 options 列表中的元素，再次执行 option.click() 就出现错误。

5.5.3 编写爬虫程序

优化程序的一个方法是每次新页面中都重新获取 <select> 元素以及下面的 <option> 元素，程序优化如下：

```
from selenium import webdriver
from selenium.webdriver.chrome.options import Options
from selenium.webdriver.common.by import By
import time

def spider():
    trs=driver.find_elements(by=By.TAG_NAME,value="tr")
    for tr in trs[1:]:
        tds=tr.find_elements(by=By.TAG_NAME,value="td")
        model=tds[0].text
        price=tds[1].text
```

```python
        print("%-16s%-16s" %(model,price))

options = Options()
options.add_argument('--headless')
options.add_argument('--disable-gpu')
driver = webdriver.Chrome(options=options)
driver.maximize_window()
driver.get("http://127.0.0.1:5000")
select = driver.find_element(by=By.ID,value="marks")
options = select.find_elements(by=By.TAG_NAME,value="option")
count=len(options)
for i in range(count):
    select = driver.find_element(by=By.ID,value="marks")
    options = select.find_elements(by=By.TAG_NAME,value="option")
    options[i].click()
    time.sleep(0.1)
    spider()
driver.close()
```

这个程序使用一个序号 i 记录当前进行的 options[i]对应页面,每次循环中都去重新获取<select>元素以及包含的所有<option>元素,保证 options[i].click()能正常执行。

5.5.4 执行爬虫程序

执行该程序爬取到了所有的产品记录如下:

MateBook D 14	5000
MateBook D 14 SE	4000
ThinkPad E14	6800
Air14Plus	3900
Dell 3400	5800

5.6 使用 Selenium 等待 HTML 元素

在浏览器加载网页的过程中,网页的加载有些元素时会有延迟的问题出现,在 HTML 元素还没有加载完成的情况下去操作这个 HTML 元素必然会出现错误,这个时候 Selenium 需要等待 HTML 元素加载完成。

5.6.1 创建延迟网站

(1) 创建网站服务器

网站服务器在访问地址时先提交 notebook.html 网页,再次访问服务器/items 时发送产品的品牌marks数据,数据按JSON字符串格式发送。为了模拟延迟过程使用time.sleep(1)

延迟1秒后发送数据,服务器server.py程序如下:

```python
import flask
import json
import time
app=flask.Flask(__name__)

@app.route("/")
def index():
    return flask.render_template("notebook.html")

@app.route("/marks")
def loadMarks():
    time.sleep(1)
    marks=["华为","联想","戴尔"]
    return json.dumps(marks)

app.debug=True
app.run()
```

(2)创建网页模板

在templates中创建一个notebook.html,这个文件使用Ajax从服务器获取产品的品牌放在<select>中。

> 注意:

<select>中的<option>开始并不存在,获取数据后才产生,模板文件如下:

```
<script>
    function loadMarks()
    {
        var http=new XMLHttpRequest();
        http.onreadystatechange=function()
        {
            if (http.readyState==4 && http.status==200)
            {
                var marks=document.getElementById("marks");
                items=eval("("+http.responseText+")");
                for(var i=0;i<items.length;i++)
                    marks.options.add(new Option(items[i],items[i]));
                document.getElementById("msg").innerHTML="品牌";
            }
        };
```

```
                    http.open("get","/marks",true);
                    http.send(null);
                }
                loadMarks();
    </script>
    <body>
      <div><span id="msg"></span><select id="marks" ></select></div>
    </body>
```

运行服务器并使用浏览器浏览结果，延迟一会后出现的网页界面，如图 5-6-1 所示。

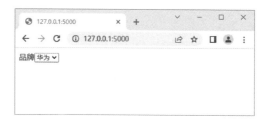

图 5-6-1
网站界面

5.6.2 编写爬虫程序

编写一个爬虫程序 spider.py 爬取产品的所有品牌，编写爬虫程序如下：

```
from selenium import webdriver
from selenium.webdriver.chrome.options import Options
from selenium.webdriver.common.by import By
import time

driver = webdriver.Chrome()
driver.get("http://127.0.0.1:5000")
options=driver.find_elements(by=By.XPATH,value="//select/option")
print("品牌数量:",len(options))
for option in options:
    print(option.text)
select=driver.find_element(by=By.XPATH,value="//select")
print(select.get_attribute("outerHTML").strip())
driver.close()
```

程序运行结果：

```
品牌数量: 0
<select id="marks"></select>
```

可见这个爬虫程序没有爬取到品牌数据，原因是服务器延迟导致这些数据还没有在网页中生成。

5.6.3 Selenium 强制等待

Selenium 使用 time.sleep(seconds)来实现强制等待 seconds 秒,这种方式是最简单的强制等待方式,不论是否可以进行下一步操作,都必须等 seconds 秒。

该方法的优点是简单,缺点是不能灵活调整需要等待的时间,有时操作还未完成,等待就结束了,导致报错。有时操作已经完成了,但等待时间还没有到,浪费时间。如果在工业中大量被使用,会浪费不必要的等待时间,影响程序执行效率。

例如,改进爬虫程序 spider.py,在加载网页后强制等待 2 秒:

```python
from selenium import webdriver
from selenium.webdriver.chrome.options import Options
from selenium.webdriver.common.by import By
import time

driver = webdriver.Chrome()
driver.get("http://127.0.0.1:5000")
time.sleep(2)
options=driver.find_elements(by=By.XPATH,value="//select/option")
print("品牌数量:",len(options))
for option in options:
    print(option.text)
select=driver.find_element(by=By.XPATH,value="//select")
print(select.get_attribute("outerHTML").strip())
driver.close()
```

程序运行结果:

```
品牌数量: 3
华为
联想
戴尔
<select id="marks"><option value="华为">华为</option><option value="联想">联想</option><option value="戴尔">戴尔</option></select>
```

由此可见,在经过等待 2 秒后程序从服务器获取了品牌数据并创建了<select>中的各个<option>元素。

5.6.4 Selenium 隐式等待

Selenium 使用 implicitly_wait(seconds)设置隐式等待指定的秒数,即网页在加载时最长等待 seconds 秒。例如,爬虫程序在访问网页设置隐式加载时间为 2 秒,编写 spider.py 程序如下:

```python
from selenium import webdriver
from selenium.webdriver.chrome.options import Options
```

```
from selenium.webdriver.common.by import By

driver = webdriver.Chrome()
#设置隐性加载时间 2 秒
driver.implicitly_wait(2)
driver.get("http://127.0.0.1:5000")
options=driver.find_elements(by=By.XPATH,value="//select/option")
print("品牌数量:",len(options))
for option in options:
    print(option.text)
select=driver.find_element(by=By.XPATH,value="//select")
print(select.get_attribute("outerHTML").strip())
driver.close()
```

执行的结果与 time.sleep(2)强制等待结果一样,都可以找到各个品牌的数据并输出。

5.6.5 Selenium 显式等待

(1) 循环等待

判断该爬虫程序能否爬到数据的关键是<select>中是否已经出现了<option>元素,可以设置一个循环来判断是否有<option>元素,修改程序 spider.py 如下:

```
from selenium import webdriver
from selenium.webdriver.chrome.options import Options
from selenium.webdriver.common.by import By
import time

driver = webdriver.Chrome()
try:
    driver.get("http://127.0.0.1:5000")
    waitTime=0
    while waitTime<10:
        print("waittime",waitTime)
        options = driver.find_elements(by=By.XPATH,value="//select/option")
        if len(options)>0:
            break
        time.sleep(0.1)
        waitTime+=0.1
        if waitTime>=10:
            raise Exception("Waiting time out")
    print("品牌数量:",len(options))
    for option in options:
        print(option.text)
```

```
            select=driver.find_element(by=By.XPATH,value="//select")
            print(select.get_attribute("outerHTML").strip())
    except Exception as err:
        print(err)
driver.close()
```

这个程序中使用 waitTime 变量来构造一个循环，最长等待约为 10 秒，每隔 0.1 秒就检查一次<select>中是否有<option>存在，如果找到了<option>元素就退出循环，不然就继续循环直到<option>出现为止或在 10 秒后还没有出现<option>就抛出异常。

程序运行结果：

```
waittime 0
waittime 0.1
waittime 0.2
waittime 0.3
waittime 0.4
waittime 0.5
waittime 0.6
waittime 0.7
waittime 0.8
waittime 0.9
品牌数量: 3
华为
联想
戴尔
<select id="marks"><option value="华为">华为</option><option value="联想">联想</option><option value="戴尔">戴尔</option></select>
```

这种循环等待的方法比强制等待更合理利用时间，不会出现<option>出现后还继续长时间等待的情况。

（2）显式等待

Selenium 的显式等待与循环等待类似。Selenium 使用 WebDriverWait 类来实现显式等待，在使用显式等待之前先引入 WebDriverWait、EC 以及 By 等类，程序如下所示：

```
from selenium.webdriver.support.wait import WebDriverWait
from selenium.webdriver.support import expected_conditions as EC
from selenium.webdriver.common.by import By
```

构造一个定位元素名为 locator 的对象。例如，通过 Xpath 的方法定位<select>中的<option>元素：

```
locator=(By.XPATH,"//select/option")
```

使用 WebDriverWait 构造一个实例，调用 until()方法：

```
WebDriverWait(driver,10, 0.1).until(EC.presence_of_element_located(locator))
```

这条语句的含义是等待 locator 指定的元素出现,最长等待 10 秒,每隔 0.1 秒就检查一次。如果在 10 秒内出现了该元素就结束等待,否则就继续等待。如果超过 10 秒还没有等待到 locator 要求的元素,就抛出一个异常。

编写显式等待的程序 spider.py 如下:

```python
from selenium import webdriver
from selenium.webdriver.chrome.options import Options
from selenium.webdriver.common.by import By
from selenium.webdriver.support.wait import WebDriverWait
from selenium.webdriver.support import expected_conditions as EC

driver = webdriver.Chrome()
try:
    driver.get("http://127.0.0.1:5000")
    locator = (By.XPATH, "//select/option")
    WebDriverWait(driver, 10,0.1).until(EC.presence_of_element_located(locator))
    options = driver.find_elements(by=By.XPATH,value="//select/option")
    print("品牌数量:",len(options))
    for option in options:
        print(option.text)
    select=driver.find_element(by=By.XPATH,value="//select")
    print(select.get_attribute("outerHTML").strip())
except Exception as err:
    print(err)
driver.close()
```

显然,程序等待到了<option>元素的出现,显式等待的优点就是等待判断相对准确,不会浪费太多等待时间,在实际中使用可以提高执行效率。

5.6.6 Selenium 等待形式

显式等待有很多种形式,读者可以查看 selenium 的文档说明,下面是一些常用的形式:

1. EC.presence_of_element_located(locator)

这种形式是等待 locator 指定的元素出现,也就是 HTML 文档中建立了这个元素。

2. EC.visibility_of_element_located(locator)

这种形式是等待 locator 指定的元素可见,注意元素出现时不一定可见,例如:

```
<select id="marks" style="display:none">...</select>
```

元素<select>出现,但不可见。

3. EC.element_to_be_clickable(locator)

这种形式是等待 locator 指定的元素可以被单击。例如,等待<option>是否可以被单击程序如下:

```
locator = (By.XPATH, "//select/option")
WebDriverWait(driver, 10,0.1).until(EC.element_to_be_clickable(locator))
```

> 注意:
>
> ```
> locator = (By.XPATH, "//select")
> WebDriverWait(driver, 10,0.1).until(EC.element_to_be_clickable(locator))
> ```
>
> 是等待<select>是否可以单击,该元素就是没有<option>时也是可以单击的,因此用这个等待爬取不到产品的品牌数据。

4. EC.element_located_to_be_selected(locator)

这种形式是等待 locator 指定的元素可以被选择,可以被选择的元素一般是<select>中的<option>、<input type="checkbox">以及<input type="radio">等元素,例如:

```
locator = (By.XPATH, "//select/option")
WebDriverWait(driver, 10,0.1).until(EC.element_located_to_be_selected(locator))
```

同样能等待加载出产品的品牌数据。

5. EC.text_to_be_present_in_element(locator,text)

这种形式是等待 locator 指定的元素的文本中包含指定的 text 文本,例如:

```
locator = (By.ID, "msg")
WebDriverWait(driver, 10,0.1).until(EC.text_to_be_present_in_element(locator,"品"))
```

等待<span id="msg">......</span>元素中包含"品"字的文本,由于在<option>出现后设置文本是"品牌",因此爬虫程序可以爬取到产品品牌数据。

总而言之,在使用 Selenium 爬取动态网页数据时,必须等待动态的网页元素加载生成完毕后才能操作这些元素。

5.7 爬取图书网站数据

本书在 1.5 节中设计了一个图书网站,现在可以使用 Selenium 设计一个爬虫程序爬取该网站的所有数据与图像。

5.7.1 网站结构分析

网站的结构分析见本书 4.5 节。

5.7.2 获取网站数据

(1)获取各个项目

图书的各个项目包含在<table>中的<tr>元素中,再循环每个<tr>元素,就可以找到每本书籍的图像与数据:

```
options = Options()
options.add_argument('--headless')
```

```
options.add_argument('--disable-gpu')
self.driver = webdriver.Chrome(options=options)
self.driver.maximize_window()
trs = self.driver.find_elements(by=By.XPATH,value="//table//tr")
for tr in trs:
    #查找每本书籍数据
```

（2）获取图书图像

<tr>中的第一个<td>就包含图片元素<img>，获取图像 src：

```
td=tr.find_element(by=By.XPATH,value="./td[position()=1]")
src=td.find_element(by=By.XPATH,value=".//img").get_attribute("src")
```

在获取图像地址 src 后设计下载图像的函数，把图像下载到 downloadBooks 文件夹中。

（3）获取图书数据

在<tr>元素最后一个<td>中包含了图书的各个数据，分析网页的结构，获取到 title（图书的名称）、author（作者）、publisher（出版社）、date（出版日期）、price（价格）等，获取方法如下：

```
td=tr.find_element(by=By.XPATH,value="./td[position()=2]")
Title=td.find_element(by=By.XPATH,value=".//div[@class='title']//h3").text
Author=td.find_element(by=By.XPATH,value=".//div[@class='author']//span[@class='attrs']").text
Publisher=td.find_element(by=By.XPATH,value=".//div[@class='publisher']//span[@class='attrs']").text
PubDate=td.find_element(by=By.XPATH,value=".//div[@class='date']//span[@class='attrs']").text
Price=td.find_element(by=By.XPATH,value=".//div[@class='price']//span[@class='price']").text
```

（4）获取换页地址

网页换页先找到<div class="paging">，再找到下面的第三个<a>元素，该元素就是下一页的超链接，单击该元素转到下一页的网址，方法如下：

```
#实现翻页
link=self.driver.find_element(by=By.XPATH,value="//div[@class='paging']/a[position()=3]")
if not link.get_attribute("href").endswith("#"):
    #找到下一页按钮，单击 click 到下一页
    link.click()
```

5.7.3 图书数据存储

爬取的数据可以存储到数据库 books.db 中，该数据库包含一张 books 表，表结构如表 4-5-1 所示。

5.7.4 编写爬虫程序

根据数据库表各个字段的要求,编写 Selenium 的爬虫程序如下:

```python
from selenium import webdriver
from selenium.webdriver.chrome.options import Options
from selenium.webdriver.common.by import By
import sqlite3
import os
import urllib.request
import time
import threading

class MySpider:
    headers = {
        "User-Agent": "Mozilla/5.0 (Windows; U; Windows NT 6.0 x64; en-US; rv: 1.9pre) Gecko/2008072421 Minefield/3.0.2pre"}
    imagePath = "downloadBooks"

    def startUp(self,url):
        # Initializing Chrome browser
        options = Options()
        options.add_argument('--headless')
        options.add_argument('--disable-gpu')
        self.driver = webdriver.Chrome(options=options)
        self.driver.maximize_window()

        # Initializing variables
        self.threads = []
        self.No = 0

        # Initializing database
        try:
            self.con=sqlite3.connect("books.db")
            self.cursor=self.con.cursor()
            try:
                self.cursor.execute("drop table books")
            except:
                pass
            sql="""
            create table books (
```

```
                    ID varchar(8) primary key,
                    Title varchar(256),
                    Author varchar(256),
                    Publisher varchar(256),
                    PubDate varchar(256),
                    Price varchar(256),
                    Ext varchar(8))
                """
            self.cursor.execute(sql)
        except Exception as err:
            print(err)

        # Initializing images folder
        try:
            if not os.path.exists(MySpider.imagePath):
                os.mkdir(MySpider.imagePath)
            images = os.listdir(MySpider.imagePath)
            for img in images:
                s = os.path.join(MySpider.imagePath, img)
                os.remove(s)
        except Exception as err:
            print(err)

        #网页第一页
        self.driver.get(url)

    def closeUp(self):
        try:
            self.con.commit()
            self.con.close()
            self.driver.close()
        except Exception as err:
            print(err)

    def insertDB(self,ID,Title,Author,Publisher,PubDate,Price,Ext):
        try:
            sql="insert into books (ID,Title,Author,Publisher,PubDate,Price,Ext) values (?,?,?,?,?,?,?)"
            self.cursor.execute(sql,[ID,Title,Author,Publisher,PubDate,Price,Ext])
        except Exception as err:
            print(err)
```

```python
def showDB(self):
    try:
        con=sqlite3.connect("books.db")
        cursor=con.cursor()
        cursor.execute("select ID,Title,Author,Publisher,PubDate,Price,Ext from books")
        rows = cursor.fetchall()
        for row in rows:
            print(row[0], row[1], row[2], row[3], row[4],row[5],row[6])
        con.close()
    except Exception as err:
        print(err)


def download(self, src,mFile):
    #下载图像
    try:
        req = urllib.request.Request(src, headers=MySpider.headers)
        resp = urllib.request.urlopen(req, timeout=400)
        data = resp.read()
        fobj = open(MySpider.imagePath + "\\" + mFile, "wb")
        fobj.write(data)
        fobj.close()
        print("downloaded ",mFile)
    except Exception as err:
        print(err)


def processSpider(self):
    #爬取一个页面的数据
    try:
        #time.sleep(1)
        #等待一会儿
        print(self.driver.current_url)
        trs = self.driver.find_elements(by=By.XPATH,value="//table//tr")
        for tr in trs:
            self.No+=1
            ID = "%06d"%self.No
            td=tr.find_element(by=By.XPATH,value="./td[position()=2]")
            Title=td.find_element(by=By.XPATH,value=".//div[@class='title']//h3").text
            Author=td.find_element(by=By.XPATH,value=".//div[@class='author']//span[@class='attrs']").text
```

```python
                    Publisher=td.find_element(by=By.XPATH,value=".//div[@class='publisher']//span[@class='attrs']").text
                    PubDate=td.find_element(by=By.XPATH,value=".//div[@class='date']//span[@class='attrs']").text
                    Price=td.find_element(by=By.XPATH,value=".//div[@class='price']//span[@class='price']").text
                    td=tr.find_element(by=By.XPATH,value="./td[position()=1]")
                    src=td.find_element(by=By.XPATH,value=".//img").get_attribute("src")
                    Ext=""
                    if src:
                        #下载图像
                        p=src.rfind(".")
                        Ext=""
                        if p>=0:
                            Ext=src[p:]
                        #启动子线程下载图像
                        mFile=ID+Ext
                        T = threading.Thread(target=self.download, args=(src,mFile))
                        T.setDaemon(False)
                        T.start()
                        self.threads.append(T)
                    self.insertDB(ID,Title,Author,Publisher,PubDate,Price,Ext)

                #实现翻页
                link=self.driver.find_element(by=By.XPATH,value="//div[@class='paging']/a[position()=3]")
                if not link.get_attribute("href").endswith("#"):
                    #找到下一页按钮，单击 click 到下一页
                    link.click()
                    self.processSpider()
        except Exception as err:
            print(err)

    def executeSpider(self, url):
        #爬取函数
        print("Spider starting......")
        self.startUp(url)
        self.processSpider()
        self.closeUp()
        #等待线程结束
        for t in self.threads:
```

```
            t.join()
            print("Spider completed......")

#主程序
url = "http://127.0.0.1:5000"
spider = MySpider()
while True:
    print("1.爬取")
    print("2.显示")
    print("3.退出")
    s=input("请选择(1,2,3):")
    if s=="1":
        spider.executeSpider(url)
    elif s=="2":
        spider.showDB()
    elif s=="3":
        break
```

执行这个爬虫程序，程序运行后爬取到了图书网站的全部图书数据与图像，结果与 3.6 节的相同。

5.8 实践项目——爬取商城网站数据

网上商城有大量的商品数据，在搜索框中输入商品名称，如"笔记本电脑"，就可以看到近百页产品的信息。使用 Selenium 编写一个爬虫程序，自动在输入框输入"笔记本电脑"，自动翻页爬取所有产品的数据与图像，并保存到数据库中。

5.8.1 解析网页代码

京东网上商城有各种各样的产品，使用浏览器进入网上商城输入"笔记本电脑"，结果，如图 5-8-1 所示。

图 5-8-1
网上商城

右击第一个产品，在快捷菜单中选择"检查"，看到每个产品的信息是包含在一组 <li> 元素中，这些 <li> 都包含在一个 <div id="J_goodsList"> 中，而且每个 <li> 的形式都是 <li class="gl-item">，因此分析 <li> 中的结构就可以找到产品的各种信息，如图 5-8-2 所示。

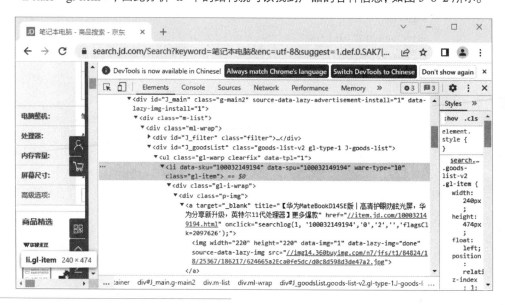

图 5-8-2
网站 HTML 代码

其中一个 <li> 的代码如下：

```
        <li class="gl-item" data-sku="100032149194" data-spu="100032149194" ware-type="10">
        <div class="gl-i-wrap">
        <div class="p-img">
        <a href="//item.jd.com/100032149194.html" onclick="searchlog(1,'100032149194','0','2','','flagsClk=2097626');" target="_blank" title="【华为 MateBookD14SE 版｜高
清护眼防眩光屏，华为分享新升级，英特尔 11 代处理器】更多爆款">
        <img data-img="1" data-lazy-img="done" height="220" source-data-lazy-img="" src="//img14.360buyimg.com/n7/jfs/t1/84824/18/25367/186217/624665a2Eca0fe5dc/d0c8d598d3de47a2.jpg" width="220"/>
        </a>
        <div data-catid="672" data-done="1" data-lease="" data-presale="0" data-venid="1000004259">
        </div>
        </div>
        <div class="p-price">
        <strong class="J_100032149194" data-done="1" data-presale="0" stock-done="1">
        <em>
         ￥
        </em>
        < i data-price="100032149194">
```

```html
      3999.00
     </i>
    </strong>
   </div>
   <div class="p-name p-name-type-2">
    <a href="//item.jd.com/100032149194.html" onclick="searchlog(1, '100032149194',
'0','1','','flagsClk=2097626');" target="_blank" title="【华为 MateBookD14SE 版丨高
     清护眼防眩光屏，华为分享新升级，英特尔 11 代处理器】更多爆款">
     <em>
      <span class="p-tag" style="background-color:#c81623">
       京品电脑
      </span>
      华为
      <font class="skcolor_ljg">
       笔记本电脑
      </font>
      MateBook D 14 SE 版 14 英寸 11 代酷睿 i5 锐炬显卡 8GB+512GB 轻薄本/高
清护眼防眩光屏银
     </em>
     <i class="promo-words" id="J_AD_100032149194">
      【华为 MateBookD14SE 版丨高清护眼防眩光屏，华为分享新升级，英特尔 11
代处理器】更多爆款
     </i>
    </a>
   </div>
   <div class="p-commit" data-done="1">
    <strong>
     <a href="//item.jd.com/100032149194.html#comment" id="J_comment_100032149194"
      onclick="searchlog(1, '100032149194','0','3','','flagsClk=2097626');" target="_blank">
      20 万+
     </a>
     条评价
    </strong>
   </div>
   <div class="p-shop" data-done="1" data-dongdong="" data-reputation="94" data-score=
"5" data-selfware="1">
    <span class="J_im_icon">
     <a class="curr-shop hd-shopname" href="//mall.jd.com/index-1000004259.html?
from=pc" onclick="searchlog(1,'1000004259',0,58)" target="_blank" title="华为京东自营官方
旗舰店">
      华为京东自营官方旗舰店
```

```html
        </a>
        <b class="im-02" onclick="searchlog(1,1000004259,0,61)" style="background:url(//img14.360buyimg.com/uba/jfs/t26764/156/1205787445/713/9f715eaa/5bc4255bN0776eea6.png) no-repeat;" title="联系客服">
        </b>
    </span>
</div>
<div class="p-icons" data-done="1" id="J_pro_100032149194">
    <i class="goods-icons J-picon-tips J-picon-fix" data-idx="1" data-tips="京东自营,品质保障">
        自营
    </i>
    <i class="goods-icons4 J-picon-tips" data-tips="本商品可领用优惠券">
        券 149-15
    </i>
</div>
<div class="p-operate">
    <a class="p-o-btn contrast J_contrast contrast" data-sku="100032149194" href="javascript:;" onclick="searchlog(1, '100032149194','0','6','','flagsClk=2097626')">    <i>
        </i>
        对比
    </a>
    <a class="p-o-btn focus J_focus" data-sku="100032149194" href="javascript:;" onclick="searchlog(1, '100032149194','0','5','','flagsClk=2097626')">
        <i>
        </i>
        关注
    </a>
    <a class="p-o-btn addcart" data-limit="0" data-stocknew="100032149194" href="//cart.jd.com/gate.action?pid=100032149194&amp;pcount=1&amp;ptype=1" onclick="searchlog(1, '100032149194','0','4','','flagsClk=2097626')" target="_blank">
        <i>
        </i>
        加入购物车
    </a>
</div>
<div class="p-stock hide" data-province="广东" data-stocknew="100032149194">
</div>
</div>
</li>
```

5.8.2 爬取网页数据

（1）爬取产品<li>元素

从 HTML 中可以看到每个<li>都包含在<div id="J_goodsList">下面的，而且每个<li>都是<li class="gl-item">的格式，程序如下：

```
lis = driver.find_elements(by=By.XPATH,value="//div[@id='J_goodsList']//li[@class='gl-item']")
for li in lis:
    # 从 li 爬取数据
```

在循环中可以得到每个<li>元素，每部产品的数据从<li>元素中进一步爬取。

（2）爬取价格

价格在<div class='p-price'>下面的一个<i>元素下，爬取价格程序如下：

```
try:
    price = li.find_element(by=By.XPATH,value=".//div[@class='p-price']//i").text
except:
    price = "0"
```

（3）爬取品牌与简介

品牌与简介在<div class='p-name p-name-type-2'>下面的<em>元素中，而且品牌是<em>下面文字中的第一个部分（用空格分开），爬取品牌 mark 与简介 note 程序如下：

```
try:
    em = li.find_element(by=By.XPATH,value=".//div[@class='p-name p-name-type- 2']//em")
    note=em.text
    mark = note.split(" ")[0]
    try:
        span=em.find_element(by=By.XPATH,value="./span[@class='p-tag']")
        mark = mark.replace(span.text+"\n", "")
        note=note.replace(span.text+"\n","")
    except:
        pass
    mark = mark.replace(",", "")
    note = note.replace(",", "")
except:
    note = ""
    mark = ""
```

（4）爬取图像地址

仔细分析 HTML 代码可以看到产品图像在<div class="p-img">下面的一个<a>超链接

的<img>元素中，图像要么存储于<img>的 src 属性中，要么存储在<img>的 data-lazy-img 属性中，因此在这个两个属性中去取两个地址 src1 与 src2，程序如下：

```
try:
    src1 = li.find_element(by=By.XPATH,value=".//div[@class='p-img']//a//img").get_attribute("src")
except:
    src1=""
try:
    src2 = li.find_element(by=By.XPATH,value=".//div[@class='p-img']//a//img").get_attribute("data-lazy-img")
except:
    src2=""

#地址 src1 与 src2 中一般总有一个有图像存在，编写 download 下载函数：
def download(self, src1, src2, mFile):
    data = None
    if src1:
        try:
            req = urllib.request.Request(src1, headers=MySpider.headers)
            resp = urllib.request.urlopen(req, timeout=400)
            data = resp.read()
        except:
            pass
    if not data and src2:
        try:
            req = urllib.request.Request(src2, headers=MySpider.headers)
            resp = urllib.request.urlopen(req, timeout=400)
            data = resp.read()
        except:
            pass
    if data:
        fobj = open(MySpider.imagePath + "\\" + mFile, "wb")
        fobj.write(data)
        fobj.close()
        print("download ", mFile)
```

如果 src1 存在 download()试图先从 src1 下载,如果 src1 不存在或者下载失败就从 src2 下载。

5.8.3 实现网页翻页

网站中产品很多，有很多个网页，找到网页的翻页控制，发现控制翻页不是通过简单的 HTML 代码控制的,而是通过 JavaScript 控制的,如图 5-8-3 所示。

图 5-8-3
网页翻页

可见要爬取下一个页面的产品就必须获取控制翻页的超链接元素<a>,并模仿鼠标单击该链接,转到下一个页面。

复制翻页的 HTML 如下:

```
<span class="p-num">
 <a class="pn-prev disabled">
  <i>
   &lt;
  </i>
  <em>
   上一页
  </em>
 </a>
 <a class="curr" href="javascript:;">
  1
 </a>
 <a href="javascript:;" onclick="SEARCH.page(3, true)">
  2
 </a>
 <a href="javascript:;" onclick="SEARCH.page(5, true)">
  3
 </a>
 <a href="javascript:;" onclick="SEARCH.page(7, true)">
  4
 </a>
```

```html
            <a href="javascript:;" onclick="SEARCH.page(9, true)">
                5
            </a>
            <a href="javascript:;" onclick="SEARCH.page(11, true)">
                6
            </a>
            <a href="javascript:;" onclick="SEARCH.page(13, true)">
                7
            </a>
            <b class="pn-break">
                ...
            </b>
            <a class="pn-next" href="javascript:;" onclick="SEARCH.page(3, true)" title="使用方向键右键也可翻到下一页！">
                <em>
                    下一页
                </em>
                <i>
                    &gt;
                </i>
            </a>
        </span>
```

只要找到<span class="p-num">，然后找到"下一页"的超链接，在正常能翻页时超链接是<a class='pn-next'>，到最后一页时变成<a class='pn-next-disabled'>，因此编写程序找到 nextPage 实现翻页：

```
try:
    self.driver.find_element (by=By.XPATH,value="//span[@class='p-num']//a[@class='pn-next disabled']")
except:
    nextPage = self.driver.find_element(by=By.XPATH,value="//span[@class='p-num']//a[@class='pn-next']")
    nextPage.click()
```

如果没有找到<a class='pn-next-disabled'>元素就找<a class='pn-next'>元素，然后使用 nextPage.click()实现翻页。

5.8.4 商品数据存储

（1）数据存储

使用 Sqlite3 数据库的一个 products.db 文件存储数据，其中有一张 items 表，表的结构如表 5-8-1 所示。

表 5-8-1　items 表结构

字段	类型	说明
mNo	varchar (8)	编号，关键字
mName	varchar (256)	产品名称
mMark	varchar (256)	产品品牌
mPrice	varchar (64)	产品价格
mNote	varchar (1024)	产品简介
mFile	varchar (256)	图像名称

下面这段程序在每次程序爬取数据前都保证建立了 items 表而且表为空：

```
self.con = sqlite3.connect("products.db")
self.cursor = self.con.cursor()
try:
    self.cursor.execute("drop table items")
except:
    pass
try:
    sql = "create table items (mNo varchar(8) primary key,mMark varchar(256),mPrice varchar(64),mNote varchar(1024),mFile varchar(32))"
    self.cursor.execute(sql)
except:
    pass
```

（2）图像存储

图像存储在 download 文件夹中，每次程序爬取图像之前保证建立了 download 文件夹，而且先删除 download 中已经存在的所有文件，程序如下：

```
if not os.path.exists(MySpider.imagePath):
    os.mkdir(MySpider.imagePath)
images = os.listdir(MySpider.imagePath)
for img in images:
    s = os.path.join(MySpider.imagePath, img)
    os.remove(s)
```

5.8.5　编写爬虫程序

根据前面的分析，编写爬虫程序 spider.py 如下：

```
from selenium import webdriver
from selenium.webdriver.chrome.options import Options
from selenium.webdriver.common.by import By
from selenium.webdriver.common.keys import Keys
```

```python
import sqlite3
import os
import urllib.request
import time
import datetime
import threading

class MySpider:
    headers = {
        "User-Agent": "Mozilla/5.0 (Windows; U; Windows NT 6.0 x64; en-US; rv:1.9pre) Gecko/2008072421 Minefield/3.0.2pre"}
    imagePath = "download"

    def startUp(self,url,key):
        # Initializing Chrome browser
        options = Options()
        options.add_argument('--headless')
        options.add_argument('--disable-gpu')
        self.driver = webdriver.Chrome(options=options)
        self.driver.maximize_window()

        # Initializing variables
        self.threads = []
        self.No = 0
        self.imgNo=0

        # Initializing database
        try:
            self.con = sqlite3.connect("products.db")
            self.cursor = self.con.cursor()
            try:
                # 如果有表就删除
                self.cursor.execute("drop table items")
            except:
                pass
            try:
                # 建立新的表
                sql = "create table items (mNo varchar(32) primary key,mMark varchar(256), mPrice varchar(32),mNote varchar(1024),mFile varchar(256))"
                self.cursor.execute(sql)
            except:
```

```python
            pass
        except Exception as err:
            print(err)


        # Initializing images folder
        try:
            if not os.path.exists(MySpider.imagePath):
                os.mkdir(MySpider.imagePath)
            images = os.listdir(MySpider.imagePath)
            for img in images:
                s = os.path.join(MySpider.imagePath, img)
                os.remove(s)
        except Exception as err:
            print(err)


        #网页第一页,输入 key 后跳转到新的页面
        self.driver.get(url)
        keyInput=self.driver.find_element(by=By.ID,value="key")
        keyInput.send_keys(key)
        keyInput.send_keys(Keys.ENTER)


    def closeUp(self):
        try:
            self.con.commit()
            self.con.close()
            self.driver.close()
        except Exception as err:
            print(err)


    def insertDB(self, mNo, mMark, mPrice, mNote, mFile):
        try:
            sql = "insert into items (mNo,mMark,mPrice,mNote,mFile) values (?,?,?,?,?)"
            self.cursor.execute(sql, (mNo, mMark, mPrice, mNote, mFile))
        except Exception as err:
            print(err)


    def showDB(self):
        try:
            con=sqlite3.connect("products.db")
            cursor=con.cursor()
            print("%-8s %-16s %-8s %-16s %s" % ("No", "Mark", "Price", "Image", "Note"))
```

```python
                cursor.execute("select mNo,mMark,mPrice,mFile,mNote from items order by mNo")
                rows = cursor.fetchall()
                for row in rows:
                    print("%-8s %-16s %-8s %-16s %s" % (row[0], row[1], row[2], row[3], row[4]))
                con.close()
            except Exception as err:
                print(err)

        def download(self, src1,src2,mFile):
            #下载图像，先从 src1 地址下载，下载不到时转到 src2 地址下载
            data=None
            if src1:
                try:
                    req = urllib.request.Request(src1, headers=MySpider.headers)
                    resp = urllib.request.urlopen(req, timeout=400)
                    data = resp.read()
                except:
                    pass
            if not data and src2:
                try:
                    req = urllib.request.Request(src2, headers=MySpider.headers)
                    resp = urllib.request.urlopen(req, timeout=400)
                    data = resp.read()
                except:
                    pass
            if data:
                fobj = open(MySpider.imagePath + "\\" + mFile, "wb")
                fobj.write(data)
                fobj.close()
                print("download ",mFile)

        def processSpider(self):
            #爬取一个页面的数据
            try:
                time.sleep(1)
                #等待一会儿
                print(self.driver.current_url)
                lis = self.driver.find_elements(by= By.XPATH,value="//div[@id='J_goodsList']//li[@class='gl-item']")
                for li in lis:
                    # We find that the image is either in src or in data-lazy-img attribute
```

```python
                    try:
                        src1 = li.find_element(by=By.XPATH,value=".//div[@class='p-img']//a//img").get_attribute("src")
                    except:
                        src1=""
                    try:
                        src2 = li.find_element(by=By.XPATH,value=".//div[@class='p-img']//a//img") .get_attribute("data-lazy-img")
                    except:
                        src2=""
                    try:
                         price = li.find_element(by=By.XPATH,value=".//div[@class='p-price']//i").text
                    except:
                        price="0"
                    try:
                        em = li.find_element(by=By.XPATH,value=".//div[@class='p-name p-name-type-2']//em")
                        note=em.text
                        mark = note.split(" ")[0]
                        try:
                            span=em.find_element(by=By.XPATH,value="./span[@class='p-tag']")
                            mark = mark.replace(span.text+"\n", "")
                            note=note.replace(span.text+"\n","")
                        except:
                            pass
                        mark = mark.replace(",", "")
                        note = note.replace(",", "")
                    except:
                        note = ""
                        mark = ""
                    self.No = self.No + 1
                    no = "%06d"%self.No
                    print(no,mark,price)
                    #爬取图像地址
                    if src1:
                        src1=urllib.request.urljoin(self.driver.current_url,src1)
                        p = src1.rfind(".")
                        mFile = no + src1[p:]
                    elif src2:
                        src2=urllib.request.urljoin(self.driver.current_url,src2)
```

```python
                        p = src2.rfind(".")
                        mFile = no + src2[p:]
                    if src1 or src2:
                        #启动子线程下载图像
                        T = threading.Thread(target=self.download, args=(src1,src2,mFile))
                        T.setDaemon(False)
                        T.start()
                        self.threads.append(T)
                    else:
                        mFile = ""
                    self.insertDB(no, mark, price, note, mFile)

                #实现翻页
                try:
                    #如果这个无效按钮存在就转到最后一页
                    self.driver.find_element(by=By.XPATH,value="//span[@class='p-num']//a[@class='pn-next disabled']")
                except:
                    #找到下一页按钮，单击 click 到下一页
                    nextPage = self.driver.find_element(by=By.XPATH,value="//span[@class='p-num']//a[@class='pn-next']")
                    nextPage.click()
                    self.processSpider()
            except Exception as err:
                print(err)


        def executeSpider(self, url,key):
            #爬取函数
            starttime = datetime.datetime.now()
            print("Spider starting......")
            self.startUp(url,key)
            self.processSpider()
            self.closeUp()
            #等待线程结束
            for t in self.threads:
                t.join()
            print("Spider completed......")
            endtime = datetime.datetime.now()
            elapsed = (endtime - starttime).seconds
            print("Total ", elapsed, " seconds elapsed")
```

```
#主程序
url = "http://www.jd.com"
spider = MySpider()
while True:
    print("1.爬取")
    print("2.显示")
    print("3.退出")
    s=input("请选择(1,2,3):")
    if s=="1":
        spider.executeSpider(url,"笔记本电脑")
    elif s=="2":
        spider.showDB()
    elif s=="3":
        break
```

startUp()函数中初始化数据库 products.db 并建立一张空的 items 表以存储数据，同时创建 download 文件夹并清空文件夹的文件，准备存储下载的图像。函数中还查找到网页的输入框<input id="key">并模拟键盘输入要爬取的商品关键字，模拟按回车键转到商品的网页，程序如下。

```
keyInput=self.driver.find_element(by=By.ID,value="key")
keyInput.send_keys(key)
keyInput.send_keys(Keys.ENTER)
```

5.8.6 执行爬虫程序

执行这个爬虫程序，结果共爬取到近 100 个页面中近 2894 条笔记本电脑的数据与图像，products.db 数据库的结果，如图 5-8-4 所示。爬取到 download 文件夹中的图像，如图 5-8-5 所示。

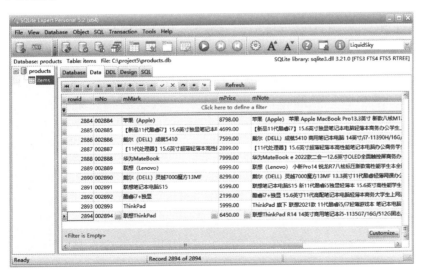

图 5-8-4 爬取到的数据

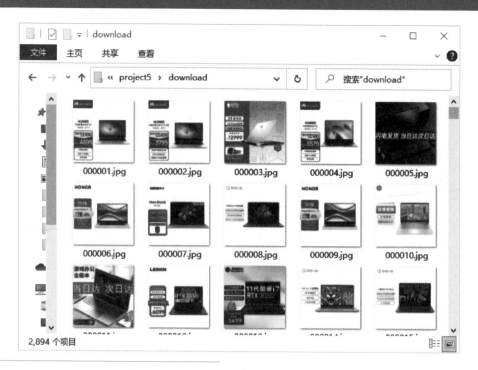

图 5-8-5
爬取到的图像

该项目爬取一个有多个页面的商城网站，使用 Selenium 技术爬取各个网页的数据，实现了爬取商城网站数据的爬虫程序。

Selenium 是一个优秀的程序框架，可以搭配多个浏览器的驱动程序，完全模拟实际的浏览器，支持静态网页与 JavaScript 控制的动态网页的数据爬取，能在程序中有效地执行 JavaScript 程序。它还支持 Xpath、CSS 等多种查找元素与数据的方法，能很好地操控各个网页元素，例如模拟键盘输入与鼠标单击等，因此使用 Selenium 能编写出功能强大的爬虫程序。

练习五

① 说明 Selenium 的工作原理，为什么能执行网页中的 JavaScript？
② 说明 Selenium 有什么等待方式，各等待方式有什么特点？
③ 启动如下的一个服务器程序。

```
import flask
app=flask.Flask(__name__)
@app.route("/")
def index():
    s='''<body>
<span id="jMsg"></span>
</body>
<script>
    document.getElementById("jMsg").innerHTML="javascript 信息";
</script>
```

```
        '''
        return s
    app.run()
```

- 用 urllib 编写程序获取网站的 HTML 代码。
- 用 Selenium 编写程序获取网站的 HTML 代码。

比较这两种方法获取的 HTML 代码是否相同，为什么？

④ 启动如下的一个服务器程序：

```
import flask
app=flask.Flask(__name__)
@app.route("/")
def index():
    s='''<body>
<span id="jMsg"></span>
</body>
<script>
   function msg()
   {
     document.getElementById("jMsg").innerHTML="javascript 信息";
   }
   window.setTimeout(msg,1000);
</script>
'''
    return s
app.run()
```

试用 Selenium 编写爬虫程序等待<span id="jMsg">...</span>中应出现的字符串，并爬取该字符串。

⑤ 启动如下的服务器程序：

```
import flask
app=flask.Flask(__name__)
@app.route("/")
def index():
    s='''<body>
<form name='frm' action='/handin' method='get'>
姓名<input type='text' name='name'><br>
性别<input type='radio' name='gender' value='男'>男<input type='radio' name='gender' value='女'>女<br>
<input type='submit' value='提交' >
</form>
'''
```

```
            return s

    @app.route("/handin")
    def handin():
        name=flask.request.values.get("name")
        gender = flask.request.values.get("gender")
        return "<div id='name'>"+name+"</div><div id='gender'>"+gender+"</div>"

    app.run()
```

试用 Selenium 编写程序：
- 填写姓名和性别是"张三"与"女"，并提交。
- 爬取提交后的姓名与性别。

结　语

　　本书通过实际案例讲解 Python 爬虫程序设计的基本方法，引领读者进入 Python 爬虫程序设计的世界。

　　编写高效的爬虫程序应该注意：

　　（1）程序的健壮性

　　爬虫程序应该有很好的健壮性，能处理爬取过程中的各种错误，不会因为出现某种错误而停止爬取。

　　（2）多线程并发

　　因为大数据的特点，爬虫程序往往要爬取成百上千万条的数据，因此爬虫程序一般是多线程的并发爬取程序，能同时爬取多个网页的数据。多线程并发的爬虫程序是一种高效的程序，其将爬取任务分解到多个不同的线程中，避免了某个爬取过程失败而导致整个爬取程序崩溃。

　　（3）优化爬取路径

　　爬虫程序往往需要在不同的网站之间来回爬取数据，控制爬虫程序爬取的深度与广度并设计一条比较优的爬取路径也是程序要考虑的重点。

　　（4）动态数据爬取

　　使用 Selenium 能爬取动态的数据，但是有些网站设置了反爬的功能，这个时候需要仔细分析网站的结构，必要时编写 JavaScript 程序改变原有网站的结构，这样才能爬取到所需数据。

　　爬虫程序功能虽然强大，但是要遵守相关的法律法规，合理爬取与使用数据。

参考文献

[1] BeautifulSoup 官方技术文档[EB/OL]. [2022-08-08]. https://www.crummy.com/software/BeautifulSoup/bs4/doc/.

[2] Python 官方技术文档[EB/OL]. [2022-08-08]. http://python.org.

[3] Scrpay 官方技术文档[EB/OL]. [2022-08-08]. https://doc.scrapy.org.

[4] Selenium 官方技术文档[EB/OL]. [2022-08-08]. https://www.selenium.dev/.

[5] 黄锐军. Python 程序设计[M]. 2 版. 北京：高等教育出版社，2021.

[6] Lawson R. 用 Python 写网络爬虫[M]. 李斌，译. 北京：人民邮电出版社，2016.

郑重声明

高等教育出版社依法对本书享有专有出版权。任何未经许可的复制、销售行为均违反《中华人民共和国著作权法》,其行为人将承担相应的民事责任和行政责任;构成犯罪的,将被依法追究刑事责任。为了维护市场秩序,保护读者的合法权益,避免读者误用盗版书造成不良后果,我社将配合行政执法部门和司法机关对违法犯罪的单位和个人进行严厉打击。社会各界人士如发现上述侵权行为,希望及时举报,我社将奖励举报有功人员。

反盗版举报电话 (010)58581999　58582371
反盗版举报邮箱 dd@hep.com.cn
通信地址 北京市西城区德外大街4号　高等教育出版社法律事务部
邮政编码 100120

读者意见反馈

为收集对教材的意见建议,进一步完善教材编写并做好服务工作,读者可将对本教材的意见建议通过如下渠道反馈至我社。

咨询电话 400-810-0598
反馈邮箱 gjdzfwb@pub.hep.cn
通信地址 北京市朝阳区惠新东街4号富盛大厦1座　高等教育出版社总编辑办公室
邮政编码 100029